CONTRACTOR'S
DAILY LOGBOOK & JOBSITE REFERENCE

ANNUAL EDITION

By Chris Prince

Published by:

DELMAR
CENGAGE Learning

www.**DeWALT**.com/guides

DELMAR
CENGAGE Learning™

DeWALT Contractor's Daily Logbook and Jobsite Reference
Chris Prince

Vice President, Technology and Trade Professional Business Unit:
Gregory L. Clayton

Product Development Manager: Robert Person

Development: Nobina Chakraborti

Director of Marketing: Beth A. Lutz

Director of Building Trades: Taryn Zlatin McKenzie

Marketing Manager: Marissa Maiella

Production Director: Carolyn Miller

Production Manager: Andrew Crouth

Art Director: Benjamin Gleeksman

For product information and technology assistance, contact us at
Cengage Learning Customer & Sales Support, 1-800-354-9706

For permission to use material from this text or product, submit all requests online at **www.cengage.com/permissions**.
Further permissions questions can be e-mailed to
permissionrequest@cengage.com

Library of Congress Control Number: 2009942557

ISBN-13: 978-1-4354-9993-5

ISBN-10: 1-4354-9993-X

Delmar
5 Maxwell Drive
Clifton Park, NY 12065-2919
USA

Cengage Learning is a leading provider of customized learning solutions with office locations around the globe, including Singapore, the United Kingdom, Australia, Mexico, Brazil, and Japan. Locate your local office at:
international.cengage.com/region

Cengage Learning products are represented in Canada by Nelson Education, Ltd.

Visit us at **www.InformationDestination.com**
For more learning solutions, please visit our corporate website at
www.cengage.com

Notice to the Reader
Cengage Learning and DeWALT do not warrant or guarantee any of the products described herein or perform any independent analysis in connection with any of the product information contained herein. Cengage Learning and DeWALT do not assume, and expressly disclaim, any obligation to obtain and include information other than that provided to it by the manufacturer. The reader is expressly warned to consider and adopt all safety precautions that might be indicated by the activities described herein and to avoid all potential hazards. By following the instructions contained herein, the reader willingly assumes all risks in connection with such instructions. Cengage Learning and DeWALT make no representations or warranties of any kind, including but not limited to, the warranties of fitness for particular purpose or merchantability, nor are any such representations implied with respect to the material set forth herein, and Cengage Learning and DeWALT take no responsibility with respect to such material. Cengage Learning and DeWALT shall not be liable for any special, consequential, or exemplary damages resulting, in whole or part, from the readers' use of, or reliance upon, this material.

Printed in China
1 2 3 4 5 6 7 12 11 10

This Book Belongs To:

Name: Normand Garand

Company: Same

Title: Owner, Carpenter

Department:

Company Address:

Company Phone: 780-719-4450

Home Phone:

Keeping a Daily Log can be one of the most important daily tasks performed by a construction manager (Superintendent, Foreman, or Project Manager; typically). By documenting the major daily activities that occur on a jobsite, the sequence of events are compiled to tell a story of how the project began, progressed and ended. If the entries are made in a consistent and diligent manner, the log can prove invaluable when the contractor's performance is questioned.

It is critical that the entries be complete. Incomplete entries can do more harm than good. A "complete" entry includes a comment that is relevant to the impact an event will have on the project. Although simple facts are important, they often do not continue the story. The fact that you "met with the architect, he provided a list of changes" is important, a comment should be added as to how the changes will impact the project. Will it postpone the completion date? If so, it should be noted in the log. For example: Met with the architect, he provided a list of changes, mostly minor with the exception of the marble selection change for the foyer. Supplier recommends we allot an extra week for delivery. Change order so notes extension of completion date. Owner has been notified.

If the fact that an inspector showed up and made an inspection is logged, the story should be completed with relevant comments. For example: Building Inspector performed rough-in inspection on Lot 107. Purlins were not properly installed. Framer has been notified and will correct tomorrow AM.

Personal opinions should never be posted to the log. It is not a forum for venting frustrations, especially those associated with internal company issues. Always remember, the log is kept for a purpose. The purpose is usually legal in nature. The contents of the log can be used to back your defense but if personal opinions and internal frustration has been documented, the log becomes possible evidence against you.

There are a few items that will be documented routinely such as weather conditions, inspections and jobsite visitors. This type of information serves many purposes. It can be used to prove delays due to weather but more importantly, it shows a consistency of verifiable documentation. If you can prove you were consistent with your log entries, your credibility is strengthened significantly and you are believed to be more trustworthy. This is very important in a court of law.

Remember, the purpose of a Daily Log is to document activities by telling the story of the project over time for the purpose of answering questions later. Think of it as a History Book that can be read about your project. But hope it is never read by anyone other than you.

Chris Prince

TODAY'S DATE	JOB/PROJECT INFORMATION	
Aug 27, 2013		
INITIALS		

HIGH TEMP									
LOW TEMP									

SCHEDULE INFORMATION	INSPECTION
Date of projected completion 28/08/13	
Is project on schedule: ☐ Yes ☒ No	
Days behind: 6 days .	

JOBSITE NOTES

Paint door, of The Parliament
Pick up locks, Architrave, Bifold HD

INJURIES			SAFETY	
Injuries on the job: ☐ Yes ☒ No			Toolbox topic:	Signage posted: ☐ Yes ☒ No
If yes, was OSHA notified: ☐ Yes ☒ No			Everyone wearing PPE: ☐ Yes ☒ No	Checklist Complete: ☒ Yes ☐ No
Type of injury: ☐ First Aid ☐ Hospital				
(*Details MUST be listed under comments*)				

TODAY'S DATE	JOB/PROJECT INFORMATION
INITIALS	

HIGH TEMP LOW TEMP									

SCHEDULE INFORMATION	INSPECTION
Date of projected completion: / /	
Is project on schedule: ☐ Yes ☐ No	
Days behind:	

JOBSITE NOTES

INJURIES

Injuries on the job: ☐ Yes ☐ No

If yes, was OSHA notified: ☐ Yes ☐ No

Type of injury: ☐ First Aid ☐ Hospital
(*Details MUST be listed under comments*)

SAFETY

Toolbox topic:	Signage posted: ☐ Yes ☐ No
Everyone wearing PPE: ☐ Yes ☐ No	Checklist Complete: ☐ Yes ☐ No

TODAY'S DATE	JOB/PROJECT INFORMATION
INITIALS	

HIGH TEMP										
LOW TEMP										

SCHEDULE INFORMATION	INSPECTION
Date of projected completion: / /	
Is project on schedule: ☐ Yes ☐ No	
Days behind:	

JOBSITE NOTES

INJURIES

Injuries on the job:	☐ Yes ☐ No
If yes, was OSHA notified:	☐ Yes ☐ No
Type of injury:	☐ First Aid ☐ Hospital

(*Details MUST be listed under comments*)

SAFETY

Toolbox topic:	Signage posted: ☐ Yes ☐ No
Everyone wearing PPE: ☐ Yes ☐ No	Checklist Complete: ☐ Yes ☐ No

TODAY'S DATE	JOB/PROJECT INFORMATION
INITIALS	

		HIGH TEMP							
HIGH TEMP									
LOW TEMP									

SCHEDULE INFORMATION	INSPECTION
Date of projected completion: / /	
Is project on schedule: ☐ Yes ☐ No	
Days behind:	

JOBSITE NOTES

INJURIES		SAFETY	
Injuries on the job: ☐ Yes ☐ No		Toolbox topic:	Signage posted: ☐ Yes ☐ No
If yes, was OSHA notified: ☐ Yes ☐ No		Everyone wearing PPE: ☐ Yes ☐ No	Checklist Complete: ☐ Yes ☐ No
Type of injury: ☐ First Aid ☐ Hospital *(Details MUST be listed under comments)*			

TODAY'S DATE	JOB/PROJECT INFORMATION
INITIALS	

HIGH TEMP LOW TEMP								

SCHEDULE INFORMATION	INSPECTION
Date of projected completion: / /	
Is project on schedule: ☐ Yes ☐ No	
Days behind:	

JOBSITE NOTES

INJURIES

Injuries on the job: ☐ Yes ☐ No	
If yes, was OSHA notified: ☐ Yes ☐ No	
Type of injury: ☐ First Aid ☐ Hospital	
(Details MUST be listed under comments)	

SAFETY

Toolbox topic:	Signage posted: ☐ Yes ☐ No
Everyone wearing PPE: ☐ Yes ☐ No	Checklist Complete: ☐ Yes ☐ No

TODAY'S DATE	JOB/PROJECT INFORMATION
INITIALS	

HIGH TEMP									
LOW TEMP									

SCHEDULE INFORMATION	INSPECTION
Date of projected completion: / /	
Is project on schedule: ☐ Yes ☐ No	
Days behind:	

JOBSITE NOTES

INJURIES	SAFETY	
Injuries on the job: ☐ Yes ☐ No	Toolbox topic:	Signage posted: ☐ Yes ☐ No
If yes, was OSHA notified: ☐ Yes ☐ No		
Type of injury: ☐ First Aid ☐ Hospital	Everyone wearing PPE: ☐ Yes ☐ No	Checklist Complete: ☐ Yes ☐ No
(Details MUST be listed under comments)		

TODAY'S DATE	JOB/PROJECT INFORMATION
INITIALS	

HIGH TEMP										
LOW TEMP										

SCHEDULE INFORMATION	INSPECTION
Date of projected completion: / /	
Is project on schedule: ☐ Yes ☐ No	
Days behind:	

JOBSITE NOTES

INJURIES		SAFETY	
Injuries on the job: ☐ Yes ☐ No		Toolbox topic:	Signage posted: ☐ Yes ☐ No
If yes, was OSHA notified: ☐ Yes ☐ No			
Type of injury: ☐ First Aid ☐ Hospital (*Details MUST be listed under comments*)		Everyone wearing PPE: ☐ Yes ☐ No	Checklist Complete: ☐ Yes ☐ No

TODAY'S DATE	JOB/PROJECT INFORMATION
INITIALS	

HIGH TEMP									
LOW TEMP									

SCHEDULE INFORMATION	INSPECTION
Date of projected completion: / /	
Is project on schedule: ☐ Yes ☐ No	
Days behind:	

JOBSITE NOTES

INJURIES	
Injuries on the job: ☐ Yes ☐ No	
If yes, was OSHA notified: ☐ Yes ☐ No	
Type of injury: ☐ First Aid ☐ Hospital	
(Details MUST be listed under comments)	

SAFETY	
Toolbox topic:	Signage posted: ☐ Yes ☐ No
Everyone wearing PPE: ☐ Yes ☐ No	Checklist Complete: ☐ Yes ☐ No

TODAY'S DATE	JOB/PROJECT INFORMATION
INITIALS	

HIGH TEMP									
LOW TEMP									

SCHEDULE INFORMATION	INSPECTION
Date of projected completion: / /	
Is project on schedule: ☐ Yes ☐ No	
Days behind:	

JOBSITE NOTES

INJURIES		SAFETY	
Injuries on the job: ☐ Yes ☐ No		Toolbox topic:	Signage posted: ☐ Yes ☐ No
If yes, was OSHA notified: ☐ Yes ☐ No			
Type of injury: ☐ First Aid ☐ Hospital		Everyone wearing PPE: ☐ Yes ☐ No	Checklist Complete: ☐ Yes ☐ No
(*Details MUST be listed under comments*)			

TODAY'S DATE	JOB/PROJECT INFORMATION
INITIALS	

HIGH TEMP		
LOW TEMP		

SCHEDULE INFORMATION	INSPECTION
Date of projected completion: / /	
Is project on schedule: ☐ Yes ☐ No	
Days behind:	

JOBSITE NOTES

INJURIES	
Injuries on the job: ☐ Yes ☐ No	
If yes, was OSHA notified: ☐ Yes ☐ No	
Type of injury: ☐ First Aid ☐ Hospital	
(Details MUST be listed under comments)	

SAFETY	
Toolbox topic:	Signage posted: ☐ Yes ☐ No
Everyone wearing PPE: ☐ Yes ☐ No	Checklist Complete: ☐ Yes ☐ No

TODAY'S DATE	JOB/PROJECT INFORMATION
INITIALS	

HIGH TEMP		
LOW TEMP		

SCHEDULE INFORMATION	INSPECTION
Date of projected completion: / /	
Is project on schedule: ☐ Yes ☐ No	
Days behind:	

JOBSITE NOTES

INJURIES
Injuries on the job: ☐ Yes   ☐ No
If yes, was OSHA notified: ☐ Yes   ☐ No
Type of injury: ☐ First Aid   ☐ Hospital
(*Details MUST be listed under comments*)

SAFETY	
Toolbox topic:	Signage posted: ☐ Yes ☐ No
Everyone wearing PPE: ☐ Yes ☐ No	Checklist Complete: ☐ Yes ☐ No

TODAY'S DATE	JOB/PROJECT INFORMATION
INITIALS	

HIGH TEMP LOW TEMP									

SCHEDULE INFORMATION	INSPECTION
Date of projected completion: / /	
Is project on schedule: ☐ Yes ☐ No	
Days behind:	

JOBSITE NOTES

INJURIES

Injuries on the job: ☐ Yes ☐ No

If yes, was OSHA notified: ☐ Yes ☐ No

Type of injury: ☐ First Aid ☐ Hospital
(Details MUST be listed under comments)

SAFETY

Toolbox topic:	Signage posted: ☐ Yes ☐ No
Everyone wearing PPE: ☐ Yes ☐ No	Checklist Complete: ☐ Yes ☐ No

TODAY'S DATE	JOB/PROJECT INFORMATION
INITIALS	

HIGH TEMP		
LOW TEMP		

SCHEDULE INFORMATION	INSPECTION
Date of projected completion: / /	
Is project on schedule: ☐ Yes ☐ No	
Days behind:	

JOBSITE NOTES

INJURIES

Injuries on the job: ☐ Yes ☐ No	
If yes, was OSHA notified: ☐ Yes ☐ No	
Type of injury: ☐ First Aid ☐ Hospital	
(*Details MUST be listed under comments*)	

SAFETY

Toolbox topic:	Signage posted: ☐ Yes ☐ No
Everyone wearing PPE: ☐ Yes ☐ No	Checklist Complete: ☐ Yes ☐ No

TODAY'S DATE	JOB/PROJECT INFORMATION
INITIALS	

HIGH TEMP									
LOW TEMP									

SCHEDULE INFORMATION	INSPECTION
Date of projected completion: / /	
Is project on schedule: ☐ Yes ☐ No	
Days behind:	

JOBSITE NOTES

INJURIES		SAFETY	
Injuries on the job: ☐ Yes ☐ No		Toolbox topic:	Signage posted: ☐ Yes ☐ No
If yes, was OSHA notified: ☐ Yes ☐ No			
Type of injury: ☐ First Aid ☐ Hospital *(Details MUST be listed under comments)*		Everyone wearing PPE: ☐ Yes ☐ No	Checklist Complete: ☐ Yes ☐ No

TODAY'S DATE	JOB/PROJECT INFORMATION
INITIALS	

HIGH TEMP									
LOW TEMP									

SCHEDULE INFORMATION	INSPECTION
Date of projected completion: / /	
Is project on schedule: ☐ Yes ☐ No	
Days behind:	

JOBSITE NOTES

_____
_____
_____
_____
_____
_____
_____
_____
_____
_____
_____
_____
_____
_____
_____
_____
_____
_____

INJURIES		SAFETY	
Injuries on the job: ☐ Yes ☐ No		Toolbox topic:	Signage posted: ☐ Yes ☐ No
If yes, was OSHA notified: ☐ Yes ☐ No			
Type of injury: ☐ First Aid ☐ Hospital		Everyone wearing PPE: ☐ Yes ☐ No	Checklist Complete: ☐ Yes ☐ No
(_Details MUST be listed under comments_)			

TODAY'S DATE	JOB/PROJECT INFORMATION
INITIALS	

HIGH TEMP									
LOW TEMP									

SCHEDULE INFORMATION	INSPECTION
Date of projected completion: / /	
Is project on schedule: ☐ Yes ☐ No	
Days behind:	

JOBSITE NOTES

INJURIES	
Injuries on the job: ☐ Yes ☐ No	
If yes, was OSHA notified: ☐ Yes ☐ No	
Type of injury: ☐ First Aid ☐ Hospital (*Details MUST be listed under comments*)	

SAFETY	
Toolbox topic:	Signage posted: ☐ Yes ☐ No
Everyone wearing PPE: ☐ Yes ☐ No	Checklist Complete: ☐ Yes ☐ No

TODAY'S DATE	JOB/PROJECT INFORMATION
INITIALS	

HIGH TEMP		
LOW TEMP		

SCHEDULE INFORMATION	INSPECTION
Date of projected completion: / /	
Is project on schedule: ☐ Yes ☐ No	
Days behind:	

JOBSITE NOTES

(blank lined notes area)

INJURIES

Injuries on the job: ☐ Yes   ☐ No

If yes, was OSHA notified: ☐ Yes   ☐ No

Type of injury: ☐ First Aid   ☐ Hospital
(*Details MUST be listed under comments*)

SAFETY

Toolbox topic:	Signage posted: ☐ Yes ☐ No
Everyone wearing PPE: ☐ Yes ☐ No	Checklist Complete: ☐ Yes ☐ No

TODAY'S DATE	JOB/PROJECT INFORMATION
INITIALS	

HIGH TEMP LOW TEMP								

SCHEDULE INFORMATION	INSPECTION
Date of projected completion: / /	
Is project on schedule: ☐ Yes ☐ No	
Days behind:	

JOBSITE NOTES

INJURIES	SAFETY	
Injuries on the job: ☐ Yes ☐ No	Toolbox topic:	Signage posted: ☐ Yes ☐ No
If yes, was OSHA notified: ☐ Yes ☐ No		
Type of injury: ☐ First Aid ☐ Hospital *(Details MUST be listed under comments)*	Everyone wearing PPE: ☐ Yes ☐ No	Checklist Complete: ☐ Yes ☐ No

TODAY'S DATE	JOB/PROJECT INFORMATION
INITIALS	

HIGH TEMP									
LOW TEMP									

SCHEDULE INFORMATION	INSPECTION
Date of projected completion: / /	
Is project on schedule: ☐ Yes ☐ No	
Days behind:	

JOBSITE NOTES

INJURIES

Injuries on the job: ☐ Yes ☐ No	
If yes, was OSHA notified: ☐ Yes ☐ No	
Type of injury: ☐ First Aid ☐ Hospital	
(*Details MUST be listed under comments*)	

SAFETY

Toolbox topic:	Signage posted: ☐ Yes ☐ No
Everyone wearing PPE: ☐ Yes ☐ No	Checklist Complete: ☐ Yes ☐ No

TODAY'S DATE	JOB/PROJECT INFORMATION
INITIALS	

HIGH TEMP										
LOW TEMP										

SCHEDULE INFORMATION	INSPECTION
Date of projected completion: / /	
Is project on schedule: ☐ Yes ☐ No	
Days behind:	

JOBSITE NOTES

INJURIES	SAFETY	
Injuries on the job: ☐ Yes ☐ No	Toolbox topic:	Signage posted: ☐ Yes ☐ No
If yes, was OSHA notified: ☐ Yes ☐ No		
Type of injury: ☐ First Aid ☐ Hospital	Everyone wearing PPE: ☐ Yes ☐ No	Checklist Complete: ☐ Yes ☐ No
(*Details MUST be listed under comments*)		

TODAY'S DATE	JOB/PROJECT INFORMATION
INITIALS	

HIGH TEMP		
LOW TEMP		

SCHEDULE INFORMATION	INSPECTION
Date of projected completion: / /	
Is project on schedule: ☐ Yes ☐ No	
Days behind:	

JOBSITE NOTES

INJURIES

Injuries on the job: ☐ Yes   ☐ No

If yes, was OSHA notified: ☐ Yes   ☐ No

Type of injury: ☐ First Aid   ☐ Hospital
(*Details MUST be listed under comments*)

SAFETY

Toolbox topic:	Signage posted: ☐ Yes ☐ No
Everyone wearing PPE: ☐ Yes ☐ No	Checklist Complete: ☐ Yes ☐ No

TODAY'S DATE	JOB/PROJECT INFORMATION
INITIALS	

HIGH TEMP LOW TEMP									

SCHEDULE INFORMATION	INSPECTION
Date of projected completion: / /	
Is project on schedule: ☐ Yes ☐ No	
Days behind:	

JOBSITE NOTES

_____

_____

_____

_____

_____

_____

_____

_____

_____

_____

_____

_____

_____

_____

_____

_____

_____

INJURIES	SAFETY	
Injuries on the job: ☐ Yes ☐ No	Toolbox topic:	Signage posted: ☐ Yes ☐ No
If yes, was OSHA notified: ☐ Yes ☐ No		
Type of injury: ☐ First Aid ☐ Hospital (_Details MUST be listed under comments_)	Everyone wearing PPE: ☐ Yes ☐ No	Checklist Complete: ☐ Yes ☐ No

TODAY'S DATE	JOB/PROJECT INFORMATION
INITIALS	

HIGH TEMP LOW TEMP								

SCHEDULE INFORMATION	INSPECTION
Date of projected completion: / /	
Is project on schedule: ☐ Yes ☐ No	
Days behind:	

JOBSITE NOTES

INJURIES
Injuries on the job: ☐ Yes   ☐ No
If yes, was OSHA notified: ☐ Yes   ☐ No
Type of injury: ☐ First Aid   ☐ Hospital
(*Details MUST be listed under comments*)

SAFETY	
Toolbox topic:	Signage posted: ☐ Yes ☐ No
Everyone wearing PPE: ☐ Yes ☐ No	Checklist Complete: ☐ Yes ☐ No

TODAY'S DATE	JOB/PROJECT INFORMATION
INITIALS	

HIGH TEMP LOW TEMP									

SCHEDULE INFORMATION	INSPECTION
Date of projected completion: / /	
Is project on schedule: ☐ Yes ☐ No	
Days behind:	

JOBSITE NOTES

_____

_____

_____

_____

_____

_____

_____

_____

_____

_____

_____

_____

_____

_____

_____

_____

INJURIES	SAFETY	
Injuries on the job: ☐ Yes ☐ No	Toolbox topic:	Signage posted: ☐ Yes ☐ No
If yes, was OSHA notified: ☐ Yes ☐ No		
Type of injury: ☐ First Aid ☐ Hospital	Everyone wearing PPE: ☐ Yes ☐ No	Checklist Complete: ☐ Yes ☐ No
(Details MUST be listed under comments)		

TODAY'S DATE	JOB/PROJECT INFORMATION
INITIALS	

HIGH TEMP										
LOW TEMP										

SCHEDULE INFORMATION	INSPECTION
Date of projected completion: / /	
Is project on schedule: ☐ Yes ☐ No	
Days behind:	

JOBSITE NOTES

INJURIES	
Injuries on the job: ☐ Yes ☐ No	
If yes, was OSHA notified: ☐ Yes ☐ No	
Type of injury: ☐ First Aid ☐ Hospital	
(*Details MUST be listed under comments*)	

SAFETY	
Toolbox topic:	Signage posted: ☐ Yes ☐ No
Everyone wearing PPE: ☐ Yes ☐ No	Checklist Complete: ☐ Yes ☐ No

TODAY'S DATE	JOB/PROJECT INFORMATION
INITIALS	

HIGH TEMP									
LOW TEMP									

SCHEDULE INFORMATION	INSPECTION
Date of projected completion: / /	
Is project on schedule: ☐ Yes ☐ No	
Days behind:	

JOBSITE NOTES

INJURIES	SAFETY	
Injuries on the job: ☐ Yes ☐ No	Toolbox topic:	Signage posted: ☐ Yes ☐ No
If yes, was OSHA notified: ☐ Yes ☐ No		
Type of injury: ☐ First Aid ☐ Hospital (*Details MUST be listed under comments*)	Everyone wearing PPE: ☐ Yes ☐ No	Checklist Complete: ☐ Yes ☐ No

TODAY'S DATE	JOB/PROJECT INFORMATION
INITIALS	

HIGH TEMP	
LOW TEMP	

SCHEDULE INFORMATION	INSPECTION
Date of projected completion: / /	
Is project on schedule: ☐ Yes ☐ No	
Days behind:	

JOBSITE NOTES

INJURIES

Injuries on the job:	☐ Yes	☐ No
If yes, was OSHA notified:	☐ Yes	☐ No
Type of injury: ☐ First Aid	☐ Hospital	

(*Details MUST be listed under comments*)

SAFETY

Toolbox topic:	Signage posted: ☐ Yes ☐ No
Everyone wearing PPE: ☐ Yes ☐ No	Checklist Complete: ☐ Yes ☐ No

TODAY'S DATE	JOB/PROJECT INFORMATION
INITIALS	

HIGH TEMP LOW TEMP									

SCHEDULE INFORMATION	INSPECTION
Date of projected completion: / /	
Is project on schedule: ☐ Yes ☐ No	
Days behind:	

JOBSITE NOTES

_____

_____

_____

_____

_____

_____

_____

_____

_____

_____

_____

_____

_____

_____

_____

_____

_____

_____

INJURIES			SAFETY	
Injuries on the job: ☐ Yes ☐ No			Toolbox topic:	Signage posted: ☐ Yes ☐ No
If yes, was OSHA notified: ☐ Yes ☐ No			Everyone wearing PPE: ☐ Yes ☐ No	Checklist Complete: ☐ Yes ☐ No
Type of injury: ☐ First Aid ☐ Hospital (*Details MUST be listed under comments*)				

TODAY'S DATE	JOB/PROJECT INFORMATION
INITIALS	

HIGH TEMP									
LOW TEMP									

SCHEDULE INFORMATION	INSPECTION
Date of projected completion: / /	
Is project on schedule: ☐ Yes ☐ No	
Days behind:	

JOBSITE NOTES

INJURIES

Injuries on the job: ☐ Yes   ☐ No

If yes, was OSHA notified: ☐ Yes   ☐ No

Type of injury: ☐ First Aid   ☐ Hospital

(*Details MUST be listed under comments*)

SAFETY

Toolbox topic:	Signage posted: ☐ Yes ☐ No
Everyone wearing PPE: ☐ Yes ☐ No	Checklist Complete: ☐ Yes ☐ No

TODAY'S DATE	JOB/PROJECT INFORMATION
INITIALS	

HIGH TEMP									
LOW TEMP									

SCHEDULE INFORMATION	INSPECTION
Date of projected completion: / /	
Is project on schedule: ☐ Yes ☐ No	
Days behind:	

JOBSITE NOTES

_____
_____
_____
_____
_____
_____
_____
_____
_____
_____
_____
_____
_____
_____
_____
_____
_____

INJURIES		SAFETY	
Injuries on the job: ☐ Yes ☐ No		Toolbox topic:	Signage posted: ☐ Yes ☐ No
If yes, was OSHA notified: ☐ Yes ☐ No		Everyone wearing PPE: ☐ Yes ☐ No	Checklist Complete: ☐ Yes ☐ No
Type of injury: ☐ First Aid ☐ Hospital _(Details MUST be listed under comments)_			

TODAY'S DATE	JOB/PROJECT INFORMATION
INITIALS	

HIGH TEMP		
LOW TEMP		

SCHEDULE INFORMATION	INSPECTION
Date of projected completion: / /	
Is project on schedule: ☐ Yes ☐ No	
Days behind:	

JOBSITE NOTES

INJURIES

Injuries on the job: ☐ Yes   ☐ No

If yes, was OSHA notified: ☐ Yes   ☐ No

Type of injury: ☐ First Aid   ☐ Hospital
(*Details MUST be listed under comments*)

SAFETY

Toolbox topic:	Signage posted: ☐ Yes ☐ No
Everyone wearing PPE: ☐ Yes ☐ No	Checklist Complete: ☐ Yes ☐ No

TODAY'S DATE	JOB/PROJECT INFORMATION
INITIALS	

HIGH TEMP									
LOW TEMP									

SCHEDULE INFORMATION	INSPECTION
Date of projected completion: / /	
Is project on schedule: ☐ Yes ☐ No	
Days behind:	

JOBSITE NOTES

INJURIES	SAFETY	
Injuries on the job: ☐ Yes ☐ No	Toolbox topic:	Signage posted: ☐ Yes ☐ No
If yes, was OSHA notified: ☐ Yes ☐ No		
Type of injury: ☐ First Aid ☐ Hospital (*Details MUST be listed under comments*)	Everyone wearing PPE: ☐ Yes ☐ No	Checklist Complete: ☐ Yes ☐ No

TODAY'S DATE	JOB/PROJECT INFORMATION
INITIALS	

HIGH TEMP									
LOW TEMP									

SCHEDULE INFORMATION	INSPECTION
Date of projected completion: / /	
Is project on schedule: ☐ Yes ☐ No	
Days behind:	

JOBSITE NOTES

INJURIES

Injuries on the job:	☐ Yes ☐ No
If yes, was OSHA notified:	☐ Yes ☐ No
Type of injury: ☐ First Aid ☐ Hospital	
(Details MUST be listed under comments)	

SAFETY

Toolbox topic:	Signage posted: ☐ Yes ☐ No
Everyone wearing PPE: ☐ Yes ☐ No	Checklist Complete: ☐ Yes ☐ No

TODAY'S DATE	JOB/PROJECT INFORMATION
INITIALS	

HIGH TEMP									
LOW TEMP									

SCHEDULE INFORMATION	INSPECTION
Date of projected completion: / /	
Is project on schedule: ☐ Yes ☐ No	
Days behind:	

JOBSITE NOTES

INJURIES		SAFETY	
Injuries on the job: ☐ Yes ☐ No		Toolbox topic:	Signage posted: ☐ Yes ☐ No
If yes, was OSHA notified: ☐ Yes ☐ No			
Type of injury: ☐ First Aid ☐ Hospital		Everyone wearing PPE: ☐ Yes ☐ No	Checklist Complete: ☐ Yes ☐ No
(Details MUST be listed under comments)			

TODAY'S DATE	JOB/PROJECT INFORMATION
INITIALS	

HIGH TEMP LOW TEMP									

SCHEDULE INFORMATION	INSPECTION
Date of projected completion: / /	
Is project on schedule: ☐ Yes ☐ No	
Days behind:	

JOBSITE NOTES

INJURIES	SAFETY	
Injuries on the job: ☐ Yes ☐ No	Toolbox topic:	Signage posted: ☐ Yes ☐ No
If yes, was OSHA notified: ☐ Yes ☐ No		
Type of injury: ☐ First Aid ☐ Hospital *(Details MUST be listed under comments)*	Everyone wearing PPE: ☐ Yes ☐ No	Checklist Complete: ☐ Yes ☐ No

TODAY'S DATE	JOB/PROJECT INFORMATION
INITIALS	

HIGH TEMP									
LOW TEMP									

SCHEDULE INFORMATION	INSPECTION
Date of projected completion: / /	
Is project on schedule: ☐ Yes ☐ No	
Days behind:	

JOBSITE NOTES

INJURIES	
Injuries on the job: ☐ Yes ☐ No	
If yes, was OSHA notified: ☐ Yes ☐ No	
Type of injury: ☐ First Aid ☐ Hospital	
(*Details MUST be listed under comments*)	

SAFETY	
Toolbox topic:	Signage posted: ☐ Yes ☐ No
Everyone wearing PPE: ☐ Yes ☐ No	Checklist Complete: ☐ Yes ☐ No

TODAY'S DATE	JOB/PROJECT INFORMATION
INITIALS	

HIGH TEMP									
LOW TEMP									

SCHEDULE INFORMATION	INSPECTION
Date of projected completion: / /	
Is project on schedule: ☐ Yes ☐ No	
Days behind:	

JOBSITE NOTES

INJURIES		SAFETY	
Injuries on the job: ☐ Yes ☐ No		Toolbox topic:	Signage posted: ☐ Yes ☐ No
If yes, was OSHA notified: ☐ Yes ☐ No			
Type of injury: ☐ First Aid ☐ Hospital (*Details MUST be listed under comments*)		Everyone wearing PPE: ☐ Yes ☐ No	Checklist Complete: ☐ Yes ☐ No

TODAY'S DATE	JOB/PROJECT INFORMATION
INITIALS	

HIGH TEMP									
LOW TEMP									

SCHEDULE INFORMATION	INSPECTION
Date of projected completion: / /	
Is project on schedule: ☐ Yes ☐ No	
Days behind:	

JOBSITE NOTES

INJURIES	SAFETY	
Injuries on the job: ☐ Yes ☐ No	Toolbox topic:	Signage posted: ☐ Yes ☐ No
If yes, was OSHA notified: ☐ Yes ☐ No		
Type of injury: ☐ First Aid ☐ Hospital	Everyone wearing PPE: ☐ Yes ☐ No	Checklist Complete: ☐ Yes ☐ No
(*Details MUST be listed under comments*)		

TODAY'S DATE	JOB/PROJECT INFORMATION
INITIALS	

HIGH TEMP								
LOW TEMP								

SCHEDULE INFORMATION	INSPECTION
Date of projected completion: / /	
Is project on schedule: ☐ Yes ☐ No	
Days behind:	

JOBSITE NOTES

INJURIES	SAFETY	
Injuries on the job: ☐ Yes ☐ No	Toolbox topic:	Signage posted: ☐ Yes ☐ No
If yes, was OSHA notified: ☐ Yes ☐ No		
Type of injury: ☐ First Aid ☐ Hospital	Everyone wearing PPE: ☐ Yes ☐ No	Checklist Complete: ☐ Yes ☐ No
(*Details MUST be listed under comments*)		

TODAY'S DATE	JOB/PROJECT INFORMATION
INITIALS	

HIGH TEMP		
LOW TEMP		

SCHEDULE INFORMATION	INSPECTION
Date of projected completion: / /	
Is project on schedule: ☐ Yes ☐ No	
Days behind:	

JOBSITE NOTES

INJURIES	
Injuries on the job: ☐ Yes ☐ No	
If yes, was OSHA notified: ☐ Yes ☐ No	
Type of injury: ☐ First Aid ☐ Hospital	
(*Details MUST be listed under comments*)	

SAFETY	
Toolbox topic:	Signage posted: ☐ Yes ☐ No
Everyone wearing PPE: ☐ Yes ☐ No	Checklist Complete: ☐ Yes ☐ No

TODAY'S DATE	JOB/PROJECT INFORMATION
INITIALS	

HIGH TEMP									
LOW TEMP									

SCHEDULE INFORMATION	INSPECTION
Date of projected completion: / /	
Is project on schedule: ☐ Yes ☐ No	
Days behind:	

JOBSITE NOTES

INJURIES			SAFETY	
Injuries on the job: ☐ Yes ☐ No			Toolbox topic:	Signage posted: ☐ Yes ☐ No
If yes, was OSHA notified: ☐ Yes ☐ No				
Type of injury: ☐ First Aid ☐ Hospital *(Details MUST be listed under comments)*			Everyone wearing PPE: ☐ Yes ☐ No	Checklist Complete: ☐ Yes ☐ No

TODAY'S DATE	JOB/PROJECT INFORMATION
INITIALS	

HIGH TEMP									
LOW TEMP									

SCHEDULE INFORMATION	INSPECTION
Date of projected completion: / /	
Is project on schedule: ☐ Yes ☐ No	
Days behind:	

JOBSITE NOTES

INJURIES

Injuries on the job: ☐ Yes   ☐ No

If yes, was OSHA notified: ☐ Yes   ☐ No

Type of injury: ☐ First Aid   ☐ Hospital
(*Details MUST be listed under comments*)

SAFETY

Toolbox topic:	Signage posted: ☐ Yes ☐ No
Everyone wearing PPE: ☐ Yes ☐ No	Checklist Complete: ☐ Yes ☐ No

TODAY'S DATE	JOB/PROJECT INFORMATION
INITIALS	

HIGH TEMP									
LOW TEMP									

SCHEDULE INFORMATION	INSPECTION
Date of projected completion: / /	
Is project on schedule: ☐ Yes ☐ No	
Days behind:	

JOBSITE NOTES

_____

_____

_____

_____

_____

_____

_____

_____

_____

_____

_____

_____

_____

_____

_____

_____

_____

INJURIES		SAFETY	
Injuries on the job: ☐ Yes ☐ No		Toolbox topic:	Signage posted: ☐ Yes ☐ No
If yes, was OSHA notified: ☐ Yes ☐ No			
Type of injury: ☐ First Aid ☐ Hospital (*Details MUST be listed under comments*)		Everyone wearing PPE: ☐ Yes ☐ No	Checklist Complete: ☐ Yes ☐ No

TODAY'S DATE	JOB/PROJECT INFORMATION
INITIALS	

HIGH TEMP										
LOW TEMP										

SCHEDULE INFORMATION	INSPECTION
Date of projected completion: / /	
Is project on schedule: ☐ Yes ☐ No	
Days behind:	

JOBSITE NOTES

INJURIES

Injuries on the job: ☐ Yes   ☐ No

If yes, was OSHA notified: ☐ Yes   ☐ No

Type of injury: ☐ First Aid   ☐ Hospital

(Details MUST be listed under comments)

SAFETY

Toolbox topic:	Signage posted: ☐ Yes ☐ No
Everyone wearing PPE: ☐ Yes ☐ No	Checklist Complete: ☐ Yes ☐ No

TODAY'S DATE	JOB/PROJECT INFORMATION
INITIALS	

HIGH TEMP LOW TEMP									

SCHEDULE INFORMATION	INSPECTION
Date of projected completion: / /	
Is project on schedule: ☐ Yes ☐ No	
Days behind:	

JOBSITE NOTES

INJURIES

Injuries on the job: ☐ Yes ☐ No	
If yes, was OSHA notified: ☐ Yes ☐ No	
Type of injury: ☐ First Aid ☐ Hospital	
(*Details MUST be listed under comments*)	

SAFETY

Toolbox topic:	Signage posted: ☐ Yes ☐ No
Everyone wearing PPE: ☐ Yes ☐ No	Checklist Complete: ☐ Yes ☐ No

TODAY'S DATE	JOB/PROJECT INFORMATION
INITIALS	

		HIGH TEMP								
HIGH TEMP										
LOW TEMP										

SCHEDULE INFORMATION	INSPECTION
Date of projected completion: / /	
Is project on schedule: ☐ Yes ☐ No	
Days behind:	

JOBSITE NOTES

INJURIES		SAFETY	
Injuries on the job: ☐ Yes ☐ No		Toolbox topic:	Signage posted: ☐ Yes ☐ No
If yes, was OSHA notified: ☐ Yes ☐ No			
Type of injury: ☐ First Aid ☐ Hospital		Everyone wearing PPE: ☐ Yes ☐ No	Checklist Complete: ☐ Yes ☐ No
(Details MUST be listed under comments)			

TODAY'S DATE	JOB/PROJECT INFORMATION
INITIALS	

HIGH TEMP LOW TEMP									

SCHEDULE INFORMATION	INSPECTION
Date of projected completion: / /	
Is project on schedule: ☐ Yes ☐ No	
Days behind:	

JOBSITE NOTES

INJURIES		
Injuries on the job: ☐ Yes ☐ No		
If yes, was OSHA notified: ☐ Yes ☐ No		
Type of injury: ☐ First Aid ☐ Hospital		
(Details MUST be listed under comments)		

SAFETY	
Toolbox topic:	Signage posted: ☐ Yes ☐ No
Everyone wearing PPE: ☐ Yes ☐ No	Checklist Complete: ☐ Yes ☐ No

TODAY'S DATE	JOB/PROJECT INFORMATION
INITIALS	

HIGH TEMP LOW TEMP								

SCHEDULE INFORMATION	INSPECTION
Date of projected completion: / /	
Is project on schedule: ☐ Yes ☐ No	
Days behind:	

JOBSITE NOTES

INJURIES		SAFETY	
Injuries on the job: ☐ Yes ☐ No		Toolbox topic:	Signage posted: ☐ Yes ☐ No
If yes, was OSHA notified: ☐ Yes ☐ No			
Type of injury: ☐ First Aid ☐ Hospital		Everyone wearing PPE: ☐ Yes ☐ No	Checklist Complete: ☐ Yes ☐ No
(*Details MUST be listed under comments*)			

TODAY'S DATE	JOB/PROJECT INFORMATION
INITIALS	

HIGH TEMP									
LOW TEMP									

SCHEDULE INFORMATION	INSPECTION
Date of projected completion: / /	
Is project on schedule: ☐ Yes ☐ No	
Days behind:	

JOBSITE NOTES

INJURIES

Injuries on the job: ☐ Yes ☐ No	
If yes, was OSHA notified: ☐ Yes ☐ No	
Type of injury: ☐ First Aid ☐ Hospital	

(*Details MUST be listed under comments*)

SAFETY

Toolbox topic:	Signage posted: ☐ Yes ☐ No
Everyone wearing PPE: ☐ Yes ☐ No	Checklist Complete: ☐ Yes ☐ No

TODAY'S DATE	JOB/PROJECT INFORMATION
INITIALS	

HIGH TEMP									
LOW TEMP									

SCHEDULE INFORMATION	INSPECTION
Date of projected completion: / /	
Is project on schedule: ☐ Yes ☐ No	
Days behind:	

JOBSITE NOTES

INJURIES

Injuries on the job: ☐ Yes   ☐ No

If yes, was OSHA notified: ☐ Yes   ☐ No

Type of injury: ☐ First Aid   ☐ Hospital

(*Details MUST be listed under comments*)

SAFETY

Toolbox topic:	Signage posted: ☐ Yes ☐ No
Everyone wearing PPE: ☐ Yes ☐ No	Checklist Complete: ☐ Yes ☐ No

TODAY'S DATE	JOB/PROJECT INFORMATION
INITIALS	

HIGH TEMP	
LOW TEMP	

SCHEDULE INFORMATION	INSPECTION
Date of projected completion: / /	
Is project on schedule: ☐ Yes ☐ No	
Days behind:	

JOBSITE NOTES

INJURIES		SAFETY	
Injuries on the job: ☐ Yes ☐ No		Toolbox topic:	Signage posted: ☐ Yes ☐ No
If yes, was OSHA notified: ☐ Yes ☐ No			
Type of injury: ☐ First Aid ☐ Hospital		Everyone wearing PPE: ☐ Yes ☐ No	Checklist Complete: ☐ Yes ☐ No
(Details MUST be listed under comments)			

TODAY'S DATE	JOB/PROJECT INFORMATION
INITIALS	

HIGH TEMP		
LOW TEMP		

SCHEDULE INFORMATION	INSPECTION
Date of projected completion: / /	
Is project on schedule: ☐ Yes ☐ No	
Days behind:	

JOBSITE NOTES

INJURIES		SAFETY	
Injuries on the job: ☐ Yes ☐ No		Toolbox topic:	Signage posted: ☐ Yes ☐ No
If yes, was OSHA notified: ☐ Yes ☐ No			
Type of injury: ☐ First Aid ☐ Hospital *(Details MUST be listed under comments)*		Everyone wearing PPE: ☐ Yes ☐ No	Checklist Complete: ☐ Yes ☐ No

TODAY'S DATE	JOB/PROJECT INFORMATION
INITIALS	

HIGH TEMP									
LOW TEMP									

SCHEDULE INFORMATION	INSPECTION
Date of projected completion: / /	
Is project on schedule: ☐ Yes ☐ No	
Days behind:	

JOBSITE NOTES

INJURIES

Injuries on the job: ☐ Yes   ☐ No
If yes, was OSHA notified: ☐ Yes   ☐ No
Type of injury: ☐ First Aid   ☐ Hospital
(*Details MUST be listed under comments*)

SAFETY

Toolbox topic:	Signage posted: ☐ Yes ☐ No
Everyone wearing PPE: ☐ Yes ☐ No	Checklist Complete: ☐ Yes ☐ No

TODAY'S DATE	JOB/PROJECT INFORMATION
INITIALS	

HIGH TEMP LOW TEMP									

SCHEDULE INFORMATION	INSPECTION
Date of projected completion: / /	
Is project on schedule: ☐ Yes ☐ No	
Days behind:	

JOBSITE NOTES

INJURIES	
Injuries on the job: ☐ Yes ☐ No	
If yes, was OSHA notified: ☐ Yes ☐ No	
Type of injury: ☐ First Aid ☐ Hospital *(Details MUST be listed under comments)*	

SAFETY	
Toolbox topic:	Signage posted: ☐ Yes ☐ No
Everyone wearing PPE: ☐ Yes ☐ No	Checklist Complete: ☐ Yes ☐ No

TODAY'S DATE	JOB/PROJECT INFORMATION
INITIALS	

HIGH TEMP		
LOW TEMP		

SCHEDULE INFORMATION	INSPECTION
Date of projected completion: / /	
Is project on schedule: ☐ Yes ☐ No	
Days behind:	

JOBSITE NOTES

(blank lined area)

INJURIES	
Injuries on the job: ☐ Yes   ☐ No
If yes, was OSHA notified: ☐ Yes   ☐ No
Type of injury: ☐ First Aid   ☐ Hospital
(*Details MUST be listed under comments*)

SAFETY	
Toolbox topic:	Signage posted: ☐ Yes ☐ No
Everyone wearing PPE: ☐ Yes ☐ No	Checklist Complete: ☐ Yes ☐ No

TODAY'S DATE	JOB/PROJECT INFORMATION
INITIALS	

HIGH TEMP		
LOW TEMP		

SCHEDULE INFORMATION	INSPECTION
Date of projected completion: / /	
Is project on schedule: ☐ Yes ☐ No	
Days behind:	

JOBSITE NOTES

INJURIES		SAFETY	
Injuries on the job: ☐ Yes ☐ No		Toolbox topic:	Signage posted: ☐ Yes ☐ No
If yes, was OSHA notified: ☐ Yes ☐ No			
Type of injury: ☐ First Aid ☐ Hospital (*Details MUST be listed under comments*)		Everyone wearing PPE: ☐ Yes ☐ No	Checklist Complete: ☐ Yes ☐ No

TODAY'S DATE	JOB/PROJECT INFORMATION
INITIALS	

HIGH TEMP									
LOW TEMP									

SCHEDULE INFORMATION	INSPECTION
Date of projected completion: / /	
Is project on schedule: ☐ Yes ☐ No	
Days behind:	

JOBSITE NOTES

INJURIES		SAFETY	
Injuries on the job: ☐ Yes ☐ No		Toolbox topic:	Signage posted: ☐ Yes ☐ No
If yes, was OSHA notified: ☐ Yes ☐ No			
Type of injury: ☐ First Aid ☐ Hospital		Everyone wearing PPE: ☐ Yes ☐ No	Checklist Complete: ☐ Yes ☐ No
(*Details MUST be listed under comments*)			

TODAY'S DATE	JOB/PROJECT INFORMATION
INITIALS	

HIGH TEMP **LOW TEMP**								

SCHEDULE INFORMATION	INSPECTION
Date of projected completion: / /	
Is project on schedule: ☐ Yes ☐ No	
Days behind:	

JOBSITE NOTES

INJURIES		SAFETY	
Injuries on the job: ☐ Yes ☐ No		Toolbox topic:	Signage posted: ☐ Yes ☐ No
If yes, was OSHA notified: ☐ Yes ☐ No			
Type of injury: ☐ First Aid ☐ Hospital		Everyone wearing PPE: ☐ Yes ☐ No	Checklist Complete: ☐ Yes ☐ No
(*Details MUST be listed under comments*)			

MAIN LOG

TODAY'S DATE	JOB/PROJECT INFORMATION
INITIALS	

HIGH TEMP LOW TEMP										

SCHEDULE INFORMATION	INSPECTION
Date of projected completion: / /	
Is project on schedule: ☐ Yes ☐ No	
Days behind:	

JOBSITE NOTES

INJURIES	SAFETY	
Injuries on the job: ☐ Yes ☐ No	Toolbox topic:	Signage posted: ☐ Yes ☐ No
If yes, was OSHA notified: ☐ Yes ☐ No		
Type of injury: ☐ First Aid ☐ Hospital (*Details MUST be listed under comments*)	Everyone wearing PPE: ☐ Yes ☐ No	Checklist Complete: ☐ Yes ☐ No

TODAY'S DATE	JOB/PROJECT INFORMATION
INITIALS	

HIGH TEMP LOW TEMP		

SCHEDULE INFORMATION	INSPECTION
Date of projected completion: / /	
Is project on schedule: ☐ Yes ☐ No	
Days behind:	

JOBSITE NOTES

INJURIES	SAFETY

INJURIES

Injuries on the job: ☐ Yes   ☐ No

If yes, was OSHA notified: ☐ Yes   ☐ No

Type of injury: ☐ First Aid   ☐ Hospital
(Details MUST be listed under comments)

SAFETY

Toolbox topic:	Signage posted: ☐ Yes ☐ No
Everyone wearing PPE: ☐ Yes ☐ No	Checklist Complete: ☐ Yes ☐ No

TODAY'S DATE	JOB/PROJECT INFORMATION
INITIALS	

HIGH TEMP LOW TEMP									

SCHEDULE INFORMATION	INSPECTION
Date of projected completion: / /	
Is project on schedule: ☐ Yes ☐ No	
Days behind:	

JOBSITE NOTES

INJURIES		SAFETY	
Injuries on the job: ☐ Yes ☐ No		Toolbox topic:	Signage posted: ☐ Yes ☐ No
If yes, was OSHA notified: ☐ Yes ☐ No			
Type of injury: ☐ First Aid ☐ Hospital		Everyone wearing PPE: ☐ Yes ☐ No	Checklist Complete: ☐ Yes ☐ No
(*Details MUST be listed under comments*)			

TODAY'S DATE	JOB/PROJECT INFORMATION
INITIALS	

HIGH TEMP LOW TEMP									

SCHEDULE INFORMATION	INSPECTION
Date of projected completion: / /	
Is project on schedule: ☐ Yes ☐ No	
Days behind:	

JOBSITE NOTES

INJURIES		SAFETY	
Injuries on the job: ☐ Yes ☐ No		Toolbox topic:	Signage posted: ☐ Yes ☐ No
If yes, was OSHA notified: ☐ Yes ☐ No			
Type of injury: ☐ First Aid ☐ Hospital (*Details MUST be listed under comments*)		Everyone wearing PPE: ☐ Yes ☐ No	Checklist Complete: ☐ Yes ☐ No

TODAY'S DATE	JOB/PROJECT INFORMATION
INITIALS	

HIGH TEMP									
LOW TEMP									

SCHEDULE INFORMATION	INSPECTION
Date of projected completion: / /	
Is project on schedule: ☐ Yes ☐ No	
Days behind:	

JOBSITE NOTES

INJURIES
Injuries on the job: ☐ Yes ☐ No
If yes, was OSHA notified: ☐ Yes ☐ No
Type of injury: ☐ First Aid ☐ Hospital
(*Details MUST be listed under comments*)

SAFETY	
Toolbox topic:	Signage posted: ☐ Yes ☐ No
Everyone wearing PPE: ☐ Yes ☐ No	Checklist Complete: ☐ Yes ☐ No

TODAY'S DATE	JOB/PROJECT INFORMATION
INITIALS	

HIGH TEMP									
LOW TEMP									

SCHEDULE INFORMATION	INSPECTION
Date of projected completion: / /	
Is project on schedule: ☐ Yes ☐ No	
Days behind:	

JOBSITE NOTES

INJURIES		SAFETY	
Injuries on the job: ☐ Yes ☐ No		Toolbox topic:	Signage posted: ☐ Yes ☐ No
If yes, was OSHA notified: ☐ Yes ☐ No			
Type of injury: ☐ First Aid ☐ Hospital (*Details MUST be listed under comments*)		Everyone wearing PPE: ☐ Yes ☐ No	Checklist Complete: ☐ Yes ☐ No

TODAY'S DATE	JOB/PROJECT INFORMATION
INITIALS	

HIGH TEMP									
LOW TEMP									

SCHEDULE INFORMATION	INSPECTION
Date of projected completion: / /	
Is project on schedule: ☐ Yes ☐ No	
Days behind:	

JOBSITE NOTES

INJURIES		SAFETY	
Injuries on the job: ☐ Yes ☐ No		Toolbox topic:	Signage posted: ☐ Yes ☐ No
If yes, was OSHA notified: ☐ Yes ☐ No			
Type of injury: ☐ First Aid ☐ Hospital		Everyone wearing PPE: ☐ Yes ☐ No	Checklist Complete: ☐ Yes ☐ No
(*Details MUST be listed under comments*)			

TODAY'S DATE	JOB/PROJECT INFORMATION
INITIALS	

HIGH TEMP									
LOW TEMP									

SCHEDULE INFORMATION	INSPECTION
Date of projected completion: / /	
Is project on schedule: ☐ Yes ☐ No	
Days behind:	

JOBSITE NOTES

INJURIES		SAFETY	
Injuries on the job: ☐ Yes ☐ No		Toolbox topic:	Signage posted: ☐ Yes ☐ No
If yes, was OSHA notified: ☐ Yes ☐ No		Everyone wearing PPE: ☐ Yes ☐ No	Checklist Complete: ☐ Yes ☐ No
Type of injury: ☐ First Aid ☐ Hospital *(Details MUST be listed under comments)*			

TODAY'S DATE	JOB/PROJECT INFORMATION
INITIALS	

HIGH TEMP								
LOW TEMP								

SCHEDULE INFORMATION	INSPECTION
Date of projected completion: / /	
Is project on schedule: ☐ Yes ☐ No	
Days behind:	

JOBSITE NOTES

INJURIES		SAFETY	
Injuries on the job: ☐ Yes ☐ No		Toolbox topic:	Signage posted: ☐ Yes ☐ No
If yes, was OSHA notified: ☐ Yes ☐ No			
Type of injury: ☐ First Aid ☐ Hospital (*Details MUST be listed under comments*)		Everyone wearing PPE: ☐ Yes ☐ No	Checklist Complete: ☐ Yes ☐ No

TODAY'S DATE	JOB/PROJECT INFORMATION
INITIALS	

HIGH TEMP									
LOW TEMP									

SCHEDULE INFORMATION	INSPECTION
Date of projected completion: / /	
Is project on schedule: ☐ Yes ☐ No	
Days behind:	

JOBSITE NOTES

INJURIES	
Injuries on the job: ☐ Yes ☐ No	
If yes, was OSHA notified: ☐ Yes ☐ No	
Type of injury: ☐ First Aid ☐ Hospital *(Details MUST be listed under comments)*	

SAFETY	
Toolbox topic:	Signage posted: ☐ Yes ☐ No
Everyone wearing PPE: ☐ Yes ☐ No	Checklist Complete: ☐ Yes ☐ No

TODAY'S DATE	JOB/PROJECT INFORMATION
INITIALS	

HIGH TEMP									
LOW TEMP									

SCHEDULE INFORMATION	INSPECTION
Date of projected completion: / /	
Is project on schedule: ☐ Yes ☐ No	
Days behind:	

JOBSITE NOTES

INJURIES		SAFETY	
Injuries on the job: ☐ Yes ☐ No		Toolbox topic:	Signage posted: ☐ Yes ☐ No
If yes, was OSHA notified: ☐ Yes ☐ No			
Type of injury: ☐ First Aid ☐ Hospital		Everyone wearing PPE: ☐ Yes ☐ No	Checklist Complete: ☐ Yes ☐ No
(*Details MUST be listed under comments*)			

TODAY'S DATE	JOB/PROJECT INFORMATION
INITIALS	

HIGH TEMP									
LOW TEMP									

SCHEDULE INFORMATION	INSPECTION
Date of projected completion: / /	
Is project on schedule: ☐ Yes ☐ No	
Days behind:	

JOBSITE NOTES

INJURIES	SAFETY	
Injuries on the job: ☐ Yes ☐ No	Toolbox topic:	Signage posted: ☐ Yes ☐ No
If yes, was OSHA notified: ☐ Yes ☐ No		
Type of injury: ☐ First Aid ☐ Hospital *(Details MUST be listed under comments)*	Everyone wearing PPE: ☐ Yes ☐ No	Checklist Complete: ☐ Yes ☐ No

TODAY'S DATE	JOB/PROJECT INFORMATION
INITIALS	

HIGH TEMP		
LOW TEMP		

SCHEDULE INFORMATION	INSPECTION
Date of projected completion: / /	
Is project on schedule: ☐ Yes ☐ No	
Days behind:	

JOBSITE NOTES

INJURIES		SAFETY	
Injuries on the job: ☐ Yes ☐ No		Toolbox topic:	Signage posted: ☐ Yes ☐ No
If yes, was OSHA notified: ☐ Yes ☐ No			
Type of injury: ☐ First Aid ☐ Hospital (*Details MUST be listed under comments*)		Everyone wearing PPE: ☐ Yes ☐ No	Checklist Complete: ☐ Yes ☐ No

TODAY'S DATE	JOB/PROJECT INFORMATION
INITIALS	

HIGH TEMP									
LOW TEMP									

SCHEDULE INFORMATION	INSPECTION
Date of projected completion: / /	
Is project on schedule: ☐ Yes ☐ No	
Days behind:	

JOBSITE NOTES

INJURIES
Injuries on the job: ☐ Yes   ☐ No
If yes, was OSHA notified: ☐ Yes   ☐ No
Type of injury: ☐ First Aid   ☐ Hospital
(*Details MUST be listed under comments*)

SAFETY	
Toolbox topic:	Signage posted: ☐ Yes ☐ No
Everyone wearing PPE: ☐ Yes ☐ No	Checklist Complete: ☐ Yes ☐ No

TODAY'S DATE	JOB/PROJECT INFORMATION
INITIALS	

HIGH TEMP								
LOW TEMP								

SCHEDULE INFORMATION	INSPECTION
Date of projected completion: / /	
Is project on schedule: ☐ Yes ☐ No	
Days behind:	

JOBSITE NOTES

INJURIES		SAFETY	
Injuries on the job: ☐ Yes ☐ No		Toolbox topic:	Signage posted: ☐ Yes ☐ No
If yes, was OSHA notified: ☐ Yes ☐ No			
Type of injury: ☐ First Aid ☐ Hospital (*Details MUST be listed under comments*)		Everyone wearing PPE: ☐ Yes ☐ No	Checklist Complete: ☐ Yes ☐ No

TODAY'S DATE	JOB/PROJECT INFORMATION
INITIALS	

HIGH TEMP		
LOW TEMP		

SCHEDULE INFORMATION	INSPECTION
Date of projected completion: / /	
Is project on schedule: ☐ Yes ☐ No	
Days behind:	

JOBSITE NOTES

INJURIES			SAFETY		
Injuries on the job: ☐ Yes ☐ No			Toolbox topic:	Signage posted: ☐ Yes ☐ No	
If yes, was OSHA notified: ☐ Yes ☐ No			Everyone wearing PPE: ☐ Yes ☐ No	Checklist Complete: ☐ Yes ☐ No	
Type of injury: ☐ First Aid ☐ Hospital *(Details MUST be listed under comments)*					

TODAY'S DATE	JOB/PROJECT INFORMATION
INITIALS	

HIGH TEMP LOW TEMP		

SCHEDULE INFORMATION	INSPECTION
Date of projected completion: / /	
Is project on schedule: ☐ Yes ☐ No	
Days behind:	

JOBSITE NOTES

INJURIES	SAFETY	
Injuries on the job: ☐ Yes ☐ No	Toolbox topic:	Signage posted: ☐ Yes ☐ No
If yes, was OSHA notified: ☐ Yes ☐ No		
Type of injury: ☐ First Aid ☐ Hospital (*Details MUST be listed under comments*)	Everyone wearing PPE: ☐ Yes ☐ No	Checklist Complete: ☐ Yes ☐ No

TODAY'S DATE	JOB/PROJECT INFORMATION
INITIALS	

HIGH TEMP **LOW TEMP**									

SCHEDULE INFORMATION / INSPECTION

SCHEDULE INFORMATION	INSPECTION
Date of projected completion: / /	
Is project on schedule: ☐ Yes ☐ No	
Days behind:	

JOBSITE NOTES

INJURIES

Injuries on the job: ☐ Yes   ☐ No

If yes, was OSHA notified: ☐ Yes   ☐ No

Type of injury: ☐ First Aid   ☐ Hospital
(Details MUST be listed under comments)

SAFETY

Toolbox topic:	Signage posted: ☐ Yes ☐ No
Everyone wearing PPE: ☐ Yes ☐ No	Checklist Complete: ☐ Yes ☐ No

TODAY'S DATE	JOB/PROJECT INFORMATION
INITIALS	

HIGH TEMP									
LOW TEMP									

SCHEDULE INFORMATION	INSPECTION
Date of projected completion: / /	
Is project on schedule: ☐ Yes ☐ No	
Days behind:	

JOBSITE NOTES

_____

_____

_____

_____

_____

_____

_____

_____

_____

_____

_____

_____

_____

_____

_____

_____

_____

_____

INJURIES

Injuries on the job: ☐ Yes ☐ No	
If yes, was OSHA notified: ☐ Yes ☐ No	
Type of injury: ☐ First Aid ☐ Hospital	
(_Details MUST be listed under comments_)	

SAFETY

Toolbox topic:	Signage posted: ☐ Yes ☐ No
Everyone wearing PPE: ☐ Yes ☐ No	Checklist Complete: ☐ Yes ☐ No

TODAY'S DATE	JOB/PROJECT INFORMATION
INITIALS	

HIGH TEMP									
LOW TEMP									

SCHEDULE INFORMATION	INSPECTION
Date of projected completion: / /	
Is project on schedule: ☐ Yes ☐ No	
Days behind:	

JOBSITE NOTES

INJURIES		SAFETY	
Injuries on the job: ☐ Yes ☐ No		Toolbox topic:	Signage posted: ☐ Yes ☐ No
If yes, was OSHA notified: ☐ Yes ☐ No			
Type of injury: ☐ First Aid ☐ Hospital		Everyone wearing PPE: ☐ Yes ☐ No	Checklist Complete: ☐ Yes ☐ No
(Details MUST be listed under comments)			

TODAY'S DATE		JOB/PROJECT INFORMATION
INITIALS		

HIGH TEMP									
LOW TEMP									

SCHEDULE INFORMATION | INSPECTION

SCHEDULE INFORMATION	INSPECTION
Date of projected completion: / /	
Is project on schedule: ☐ Yes ☐ No	
Days behind:	

JOBSITE NOTES

INJURIES

INJURIES
Injuries on the job: ☐ Yes   ☐ No
If yes, was OSHA notified: ☐ Yes   ☐ No
Type of injury: ☐ First Aid   ☐ Hospital
(Details MUST be listed under comments)

SAFETY

Toolbox topic:	Signage posted: ☐ Yes ☐ No
Everyone wearing PPE: ☐ Yes ☐ No	Checklist Complete: ☐ Yes ☐ No

TODAY'S DATE	JOB/PROJECT INFORMATION
INITIALS	

HIGH TEMP LOW TEMP									

SCHEDULE INFORMATION	INSPECTION
Date of projected completion: / /	
Is project on schedule: ☐ Yes ☐ No	
Days behind:	

JOBSITE NOTES

INJURIES		SAFETY	
Injuries on the job: ☐ Yes ☐ No		Toolbox topic:	Signage posted: ☐ Yes ☐ No
If yes, was OSHA notified: ☐ Yes ☐ No			
Type of injury: ☐ First Aid ☐ Hospital		Everyone wearing PPE: ☐ Yes ☐ No	Checklist Complete: ☐ Yes ☐ No
(*Details MUST be listed under comments*)			

TODAY'S DATE	JOB/PROJECT INFORMATION
INITIALS	

HIGH TEMP									
LOW TEMP									

SCHEDULE INFORMATION	INSPECTION
Date of projected completion: / /	
Is project on schedule: ☐ Yes ☐ No	
Days behind:	

JOBSITE NOTES

INJURIES		SAFETY	
Injuries on the job: ☐ Yes ☐ No		Toolbox topic:	Signage posted: ☐ Yes ☐ No
If yes, was OSHA notified: ☐ Yes ☐ No			
Type of injury: ☐ First Aid ☐ Hospital (*Details MUST be listed under comments*)		Everyone wearing PPE: ☐ Yes ☐ No	Checklist Complete: ☐ Yes ☐ No

TODAY'S DATE	JOB/PROJECT INFORMATION
INITIALS	

HIGH TEMP LOW TEMP									

SCHEDULE INFORMATION	INSPECTION
Date of projected completion: / /	
Is project on schedule: ☐ Yes ☐ No	
Days behind:	

JOBSITE NOTES

_____
_____
_____
_____
_____
_____
_____
_____
_____
_____
_____
_____
_____
_____
_____

INJURIES

Injuries on the job: ☐ Yes ☐ No

If yes, was OSHA notified: ☐ Yes ☐ No

Type of injury: ☐ First Aid ☐ Hospital
(*Details MUST be listed under comments*)

SAFETY

Toolbox topic:	Signage posted: ☐ Yes ☐ No
Everyone wearing PPE: ☐ Yes ☐ No	Checklist Complete: ☐ Yes ☐ No

TODAY'S DATE	JOB/PROJECT INFORMATION
INITIALS	

HIGH TEMP **LOW TEMP**								

SCHEDULE INFORMATION	INSPECTION
Date of projected completion: / /	
Is project on schedule: ☐ Yes ☐ No	
Days behind:	

JOBSITE NOTES

INJURIES	SAFETY	
Injuries on the job: ☐ Yes ☐ No	Toolbox topic:	Signage posted: ☐ Yes ☐ No
If yes, was OSHA notified: ☐ Yes ☐ No		
Type of injury: ☐ First Aid ☐ Hospital (*Details MUST be listed under comments*)	Everyone wearing PPE: ☐ Yes ☐ No	Checklist Complete: ☐ Yes ☐ No

TODAY'S DATE	JOB/PROJECT INFORMATION
INITIALS	

HIGH TEMP									
LOW TEMP									

SCHEDULE INFORMATION	INSPECTION
Date of projected completion: / /	
Is project on schedule: ☐ Yes ☐ No	
Days behind:	

JOBSITE NOTES

INJURIES		SAFETY	
Injuries on the job: ☐ Yes ☐ No		Toolbox topic:	Signage posted: ☐ Yes ☐ No
If yes, was OSHA notified: ☐ Yes ☐ No			
Type of injury: ☐ First Aid ☐ Hospital (*Details MUST be listed under comments*)		Everyone wearing PPE: ☐ Yes ☐ No	Checklist Complete: ☐ Yes ☐ No

TODAY'S DATE	JOB/PROJECT INFORMATION
INITIALS	

HIGH TEMP									
LOW TEMP									

SCHEDULE INFORMATION	INSPECTION
Date of projected completion: / /	
Is project on schedule: ☐ Yes ☐ No	
Days behind:	

JOBSITE NOTES

INJURIES		SAFETY	
Injuries on the job: ☐ Yes ☐ No		Toolbox topic:	Signage posted: ☐ Yes ☐ No
If yes, was OSHA notified: ☐ Yes ☐ No			
Type of injury: ☐ First Aid ☐ Hospital		Everyone wearing PPE: ☐ Yes ☐ No	Checklist Complete: ☐ Yes ☐ No
(Details MUST be listed under comments)			

TODAY'S DATE	JOB/PROJECT INFORMATION
INITIALS	

HIGH TEMP LOW TEMP									

SCHEDULE INFORMATION	INSPECTION
Date of projected completion: / /	
Is project on schedule: ☐ Yes ☐ No	
Days behind:	

JOBSITE NOTES

INJURIES			SAFETY	
Injuries on the job: ☐ Yes ☐ No			Toolbox topic:	Signage posted: ☐ Yes ☐ No
If yes, was OSHA notified: ☐ Yes ☐ No			Everyone wearing PPE: ☐ Yes ☐ No	Checklist Complete: ☐ Yes ☐ No
Type of injury: ☐ First Aid ☐ Hospital *(Details MUST be listed under comments)*				

TODAY'S DATE	JOB/PROJECT INFORMATION
INITIALS	

HIGH TEMP									
LOW TEMP									

SCHEDULE INFORMATION	INSPECTION
Date of projected completion: / /	
Is project on schedule: ☐ Yes ☐ No	
Days behind:	

JOBSITE NOTES

_____
_____
_____
_____
_____
_____
_____
_____
_____
_____
_____
_____
_____
_____
_____
_____
_____
_____
_____

INJURIES	
Injuries on the job: ☐ Yes ☐ No	
If yes, was OSHA notified: ☐ Yes ☐ No	
Type of injury: ☐ First Aid ☐ Hospital	
(_Details MUST be listed under comments_)	

SAFETY	
Toolbox topic:	Signage posted: ☐ Yes ☐ No
Everyone wearing PPE: ☐ Yes ☐ No	Checklist Complete: ☐ Yes ☐ No

TODAY'S DATE	JOB/PROJECT INFORMATION
INITIALS	

HIGH TEMP									
LOW TEMP									

SCHEDULE INFORMATION	INSPECTION
Date of projected completion: / /	
Is project on schedule: ☐ Yes ☐ No	
Days behind:	

JOBSITE NOTES

INJURIES		SAFETY	
Injuries on the job: ☐ Yes ☐ No		Toolbox topic:	Signage posted: ☐ Yes ☐ No
If yes, was OSHA notified: ☐ Yes ☐ No			
Type of injury: ☐ First Aid ☐ Hospital	Everyone wearing PPE: ☐ Yes ☐ No	Checklist Complete: ☐ Yes ☐ No
(Details MUST be listed under comments)			

TODAY'S DATE	JOB/PROJECT INFORMATION
INITIALS	

HIGH TEMP LOW TEMP									

SCHEDULE INFORMATION	INSPECTION
Date of projected completion: / /	
Is project on schedule: ☐ Yes ☐ No	
Days behind:	

JOBSITE NOTES

INJURIES

Injuries on the job: ☐ Yes ☐ No

If yes, was OSHA notified: ☐ Yes ☐ No

Type of injury: ☐ First Aid ☐ Hospital
(Details MUST be listed under comments)

SAFETY

Toolbox topic:	Signage posted: ☐ Yes ☐ No
Everyone wearing PPE: ☐ Yes ☐ No	Checklist Complete: ☐ Yes ☐ No

TODAY'S DATE	JOB/PROJECT INFORMATION
INITIALS	

HIGH TEMP		
LOW TEMP		

SCHEDULE INFORMATION	INSPECTION
Date of projected completion: / /	
Is project on schedule: ☐ Yes ☐ No	
Days behind:	

JOBSITE NOTES

INJURIES

Injuries on the job: ☐ Yes ☐ No	
If yes, was OSHA notified: ☐ Yes ☐ No	
Type of injury: ☐ First Aid ☐ Hospital	
(*Details MUST be listed under comments*)	

SAFETY

Toolbox topic:	Signage posted: ☐ Yes ☐ No
Everyone wearing PPE: ☐ Yes ☐ No	Checklist Complete: ☐ Yes ☐ No

TODAY'S DATE	JOB/PROJECT INFORMATION
INITIALS	

HIGH TEMP									
LOW TEMP									

SCHEDULE INFORMATION	INSPECTION
Date of projected completion: / /	
Is project on schedule: ☐ Yes ☐ No	
Days behind:	

JOBSITE NOTES

INJURIES	SAFETY	
Injuries on the job: ☐ Yes ☐ No	Toolbox topic:	Signage posted: ☐ Yes ☐ No
If yes, was OSHA notified: ☐ Yes ☐ No		
Type of injury: ☐ First Aid ☐ Hospital	Everyone wearing PPE: ☐ Yes ☐ No	Checklist Complete: ☐ Yes ☐ No
(*Details MUST be listed under comments*)		

TODAY'S DATE	JOB/PROJECT INFORMATION
INITIALS	

HIGH TEMP			
LOW TEMP			

SCHEDULE INFORMATION	INSPECTION
Date of projected completion: / /	
Is project on schedule: ☐ Yes ☐ No	
Days behind:	

JOBSITE NOTES

INJURIES		SAFETY	
Injuries on the job: ☐ Yes ☐ No		Toolbox topic:	Signage posted: ☐ Yes ☐ No
If yes, was OSHA notified: ☐ Yes ☐ No			
Type of injury: ☐ First Aid ☐ Hospital		Everyone wearing PPE: ☐ Yes ☐ No	Checklist Complete: ☐ Yes ☐ No
(*Details MUST be listed under comments*)			

TODAY'S DATE	JOB/PROJECT INFORMATION
INITIALS	

HIGH TEMP									
LOW TEMP									

SCHEDULE INFORMATION / INSPECTION

Date of projected completion: / /

Is project on schedule: ☐ Yes ☐ No

Days behind:

JOBSITE NOTES

INJURIES

Injuries on the job: ☐ Yes ☐ No

If yes, was OSHA notified: ☐ Yes ☐ No

Type of injury: ☐ First Aid ☐ Hospital
(*Details MUST be listed under comments*)

SAFETY

Toolbox topic:	Signage posted: ☐ Yes ☐ No
Everyone wearing PPE: ☐ Yes ☐ No	Checklist Complete: ☐ Yes ☐ No

MAIN LOG

TODAY'S DATE	JOB/PROJECT INFORMATION
INITIALS	

HIGH TEMP LOW TEMP								

SCHEDULE INFORMATION	INSPECTION
Date of projected completion: / /	
Is project on schedule: ☐ Yes ☐ No	
Days behind:	

JOBSITE NOTES

INJURIES
Injuries on the job: ☐ Yes   ☐ No
If yes, was OSHA notified: ☐ Yes   ☐ No
Type of injury: ☐ First Aid   ☐ Hospital
(*Details MUST be listed under comments*)

SAFETY	
Toolbox topic:	Signage posted: ☐ Yes ☐ No
Everyone wearing PPE: ☐ Yes ☐ No	Checklist Complete: ☐ Yes ☐ No

TODAY'S DATE	JOB/PROJECT INFORMATION
INITIALS	

HIGH TEMP									
LOW TEMP									

SCHEDULE INFORMATION	INSPECTION
Date of projected completion: / /	
Is project on schedule: ☐ Yes ☐ No	
Days behind:	

JOBSITE NOTES

INJURIES		SAFETY	
Injuries on the job: ☐ Yes ☐ No		Toolbox topic:	Signage posted: ☐ Yes ☐ No
If yes, was OSHA notified: ☐ Yes ☐ No			
Type of injury: ☐ First Aid ☐ Hospital		Everyone wearing PPE: ☐ Yes ☐ No	Checklist Complete: ☐ Yes ☐ No
(*Details MUST be listed under comments*)			

TODAY'S DATE	JOB/PROJECT INFORMATION
INITIALS	

HIGH TEMP LOW TEMP								

SCHEDULE INFORMATION	INSPECTION
Date of projected completion: / /	
Is project on schedule: ☐ Yes ☐ No	
Days behind:	

JOBSITE NOTES

_____

_____

_____

_____

_____

_____

_____

_____

_____

_____

_____

_____

_____

_____

_____

_____

INJURIES	SAFETY	
Injuries on the job: ☐ Yes ☐ No	Toolbox topic:	Signage posted: ☐ Yes ☐ No
If yes, was OSHA notified: ☐ Yes ☐ No	Everyone wearing PPE: ☐ Yes ☐ No	Checklist Complete: ☐ Yes ☐ No
Type of injury: ☐ First Aid ☐ Hospital _(Details MUST be listed under comments)_		

TODAY'S DATE	JOB/PROJECT INFORMATION
INITIALS	

HIGH TEMP		
LOW TEMP		

SCHEDULE INFORMATION	INSPECTION
Date of projected completion: / /	
Is project on schedule: ☐ Yes ☐ No	
Days behind:	

JOBSITE NOTES

INJURIES	
Injuries on the job: ☐ Yes ☐ No	
If yes, was OSHA notified: ☐ Yes ☐ No	
Type of injury: ☐ First Aid ☐ Hospital	
(*Details MUST be listed under comments*)	

SAFETY	
Toolbox topic:	Signage posted: ☐ Yes ☐ No
Everyone wearing PPE: ☐ Yes ☐ No	Checklist Complete: ☐ Yes ☐ No

TODAY'S DATE	JOB/PROJECT INFORMATION
INITIALS	

HIGH TEMP								
LOW TEMP								

SCHEDULE INFORMATION	INSPECTION
Date of projected completion: / /	
Is project on schedule: ☐ Yes ☐ No	
Days behind:	

JOBSITE NOTES

INJURIES		SAFETY	
Injuries on the job: ☐ Yes ☐ No		Toolbox topic:	Signage posted: ☐ Yes ☐ No
If yes, was OSHA notified: ☐ Yes ☐ No			
Type of injury: ☐ First Aid ☐ Hospital		Everyone wearing PPE: ☐ Yes ☐ No	Checklist Complete: ☐ Yes ☐ No
(Details MUST be listed under comments)			

TODAY'S DATE	JOB/PROJECT INFORMATION
INITIALS	

HIGH TEMP									
LOW TEMP									

SCHEDULE INFORMATION	INSPECTION
Date of projected completion: / /	
Is project on schedule: ☐ Yes ☐ No	
Days behind:	

JOBSITE NOTES

INJURIES
Injuries on the job: ☐ Yes ☐ No
If yes, was OSHA notified: ☐ Yes ☐ No
Type of injury: ☐ First Aid ☐ Hospital
(*Details MUST be listed under comments*)

SAFETY	
Toolbox topic:	Signage posted: ☐ Yes ☐ No
Everyone wearing PPE: ☐ Yes ☐ No	Checklist Complete: ☐ Yes ☐ No

TODAY'S DATE	JOB/PROJECT INFORMATION
INITIALS	

HIGH TEMP									
LOW TEMP									

SCHEDULE INFORMATION	INSPECTION
Date of projected completion: / /	
Is project on schedule: ☐ Yes ☐ No	
Days behind:	

JOBSITE NOTES

INJURIES

Injuries on the job: ☐ Yes   ☐ No

If yes, was OSHA notified: ☐ Yes   ☐ No

Type of injury: ☐ First Aid   ☐ Hospital
(*Details MUST be listed under comments*)

SAFETY

Toolbox topic:	Signage posted: ☐ Yes ☐ No
Everyone wearing PPE: ☐ Yes ☐ No	Checklist Complete: ☐ Yes ☐ No

TODAY'S DATE	JOB/PROJECT INFORMATION
INITIALS	

HIGH TEMP									
LOW TEMP									

SCHEDULE INFORMATION	INSPECTION
Date of projected completion: / /	
Is project on schedule: ☐ Yes ☐ No	
Days behind:	

JOBSITE NOTES

INJURIES	SAFETY	
Injuries on the job: ☐ Yes ☐ No	Toolbox topic:	Signage posted: ☐ Yes ☐ No
If yes, was OSHA notified: ☐ Yes ☐ No		
Type of injury: ☐ First Aid ☐ Hospital	Everyone wearing PPE: ☐ Yes ☐ No	Checklist Complete: ☐ Yes ☐ No
(*Details MUST be listed under comments*)		

TODAY'S DATE	JOB/PROJECT INFORMATION
INITIALS	

HIGH TEMP									
LOW TEMP									

SCHEDULE INFORMATION	INSPECTION
Date of projected completion: / /	
Is project on schedule: ☐ Yes ☐ No	
Days behind:	

JOBSITE NOTES

INJURIES		SAFETY	
Injuries on the job: ☐ Yes ☐ No		Toolbox topic:	Signage posted: ☐ Yes ☐ No
If yes, was OSHA notified: ☐ Yes ☐ No			
Type of injury: ☐ First Aid ☐ Hospital		Everyone wearing PPE: ☐ Yes ☐ No	Checklist Complete: ☐ Yes ☐ No
(*Details MUST be listed under comments*)			

TODAY'S DATE	JOB/PROJECT INFORMATION
INITIALS	

HIGH TEMP LOW TEMP									

SCHEDULE INFORMATION	INSPECTION
Date of projected completion: / /	
Is project on schedule: ☐ Yes ☐ No	
Days behind:	

JOBSITE NOTES

INJURIES	
Injuries on the job: ☐ Yes ☐ No	
If yes, was OSHA notified: ☐ Yes ☐ No	
Type of injury: ☐ First Aid ☐ Hospital	
(*Details MUST be listed under comments*)	

SAFETY	
Toolbox topic:	Signage posted: ☐ Yes ☐ No
Everyone wearing PPE: ☐ Yes ☐ No	Checklist Complete: ☐ Yes ☐ No

TODAY'S DATE	JOB/PROJECT INFORMATION
INITIALS	

HIGH TEMP		
LOW TEMP		

SCHEDULE INFORMATION	INSPECTION
Date of projected completion: / /	
Is project on schedule: ☐ Yes ☐ No	
Days behind:	

JOBSITE NOTES

INJURIES

Injuries on the job:	☐ Yes	☐ No
If yes, was OSHA notified:	☐ Yes	☐ No
Type of injury:	☐ First Aid	☐ Hospital

(*Details MUST be listed under comments*)

SAFETY

Toolbox topic:	Signage posted: ☐ Yes ☐ No
Everyone wearing PPE: ☐ Yes ☐ No	Checklist Complete: ☐ Yes ☐ No

TODAY'S DATE	JOB/PROJECT INFORMATION
INITIALS	

HIGH TEMP									
LOW TEMP									

SCHEDULE INFORMATION	INSPECTION
Date of projected completion: / /	
Is project on schedule: ☐ Yes ☐ No	
Days behind:	

JOBSITE NOTES

INJURIES		SAFETY	
Injuries on the job: ☐ Yes ☐ No		Toolbox topic:	Signage posted: ☐ Yes ☐ No
If yes, was OSHA notified: ☐ Yes ☐ No			
Type of injury: ☐ First Aid ☐ Hospital		Everyone wearing PPE: ☐ Yes ☐ No	Checklist Complete: ☐ Yes ☐ No
(*Details MUST be listed under comments*)			

TODAY'S DATE	JOB/PROJECT INFORMATION
INITIALS	

HIGH TEMP									
LOW TEMP									

SCHEDULE INFORMATION	INSPECTION
Date of projected completion: / /	
Is project on schedule: ☐ Yes ☐ No	
Days behind:	

JOBSITE NOTES

INJURIES	SAFETY	
Injuries on the job: ☐ Yes ☐ No	Toolbox topic:	Signage posted: ☐ Yes ☐ No
If yes, was OSHA notified: ☐ Yes ☐ No		
Type of injury: ☐ First Aid ☐ Hospital *(Details MUST be listed under comments)*	Everyone wearing PPE: ☐ Yes ☐ No	Checklist Complete: ☐ Yes ☐ No

TODAY'S DATE	JOB/PROJECT INFORMATION
INITIALS	

HIGH TEMP		
LOW TEMP		

SCHEDULE INFORMATION	INSPECTION
Date of projected completion: / /	
Is project on schedule: ☐ Yes ☐ No	
Days behind:	

JOBSITE NOTES

INJURIES	SAFETY	
Injuries on the job: ☐ Yes ☐ No	Toolbox topic:	Signage posted: ☐ Yes ☐ No
If yes, was OSHA notified: ☐ Yes ☐ No		
Type of injury: ☐ First Aid ☐ Hospital (*Details MUST be listed under comments*)	Everyone wearing PPE: ☐ Yes ☐ No	Checklist Complete: ☐ Yes ☐ No

TODAY'S DATE	JOB/PROJECT INFORMATION
INITIALS	

		HIGH TEMP								

HIGH TEMP

LOW TEMP

SCHEDULE INFORMATION	INSPECTION
Date of projected completion: / /	
Is project on schedule: ☐ Yes ☐ No	
Days behind:	

JOBSITE NOTES

INJURIES			SAFETY	
Injuries on the job: ☐ Yes ☐ No			Toolbox topic:	Signage posted: ☐ Yes ☐ No
If yes, was OSHA notified: ☐ Yes ☐ No			Everyone wearing PPE: ☐ Yes ☐ No	Checklist Complete: ☐ Yes ☐ No
Type of injury: ☐ First Aid ☐ Hospital *(Details MUST be listed under comments)*				

TODAY'S DATE	JOB/PROJECT INFORMATION
INITIALS	

HIGH TEMP								
LOW TEMP								

SCHEDULE INFORMATION	INSPECTION
Date of projected completion: / /	
Is project on schedule: ☐ Yes ☐ No	
Days behind:	

JOBSITE NOTES

INJURIES		SAFETY	
Injuries on the job: ☐ Yes ☐ No		Toolbox topic:	Signage posted: ☐ Yes ☐ No
If yes, was OSHA notified: ☐ Yes ☐ No			
Type of injury: ☐ First Aid ☐ Hospital *(Details MUST be listed under comments)*		Everyone wearing PPE: ☐ Yes ☐ No	Checklist Complete: ☐ Yes ☐ No

TODAY'S DATE	JOB/PROJECT INFORMATION
INITIALS	

HIGH TEMP									
LOW TEMP									

SCHEDULE INFORMATION	INSPECTION
Date of projected completion: / /	
Is project on schedule: ☐ Yes ☐ No	
Days behind:	

JOBSITE NOTES

_____

_____

_____

_____

_____

_____

_____

_____

_____

_____

_____

_____

_____

_____

_____

_____

INJURIES		SAFETY	
Injuries on the job: ☐ Yes ☐ No		Toolbox topic:	Signage posted: ☐ Yes ☐ No
If yes, was OSHA notified: ☐ Yes ☐ No		Everyone wearing PPE: ☐ Yes ☐ No	Checklist Complete: ☐ Yes ☐ No
Type of injury: ☐ First Aid ☐ Hospital			
(Details MUST be listed under comments)			

TODAY'S DATE	JOB/PROJECT INFORMATION
INITIALS	

HIGH TEMP									
LOW TEMP									

SCHEDULE INFORMATION	INSPECTION
Date of projected completion: / /	
Is project on schedule: ☐ Yes ☐ No	
Days behind:	

JOBSITE NOTES

_____
_____
_____
_____
_____
_____
_____
_____
_____
_____
_____
_____
_____
_____
_____
_____
_____

INJURIES

Injuries on the job: ☐ Yes ☐ No

If yes, was OSHA notified: ☐ Yes ☐ No

Type of injury: ☐ First Aid ☐ Hospital
(*Details MUST be listed under comments*)

SAFETY

Toolbox topic:	Signage posted: ☐ Yes ☐ No
Everyone wearing PPE: ☐ Yes ☐ No	Checklist Complete: ☐ Yes ☐ No

TODAY'S DATE	JOB/PROJECT INFORMATION
INITIALS	

HIGH TEMP									
LOW TEMP									

SCHEDULE INFORMATION	INSPECTION
Date of projected completion: / /	
Is project on schedule: ☐ Yes ☐ No	
Days behind:	

JOBSITE NOTES

_____

_____

_____

_____

_____

_____

_____

_____

_____

_____

_____

_____

_____

_____

_____

INJURIES	SAFETY	
Injuries on the job: ☐ Yes ☐ No	Toolbox topic:	Signage posted: ☐ Yes ☐ No
If yes, was OSHA notified: ☐ Yes ☐ No		
Type of injury: ☐ First Aid ☐ Hospital	Everyone wearing PPE: ☐ Yes ☐ No	Checklist Complete: ☐ Yes ☐ No
(Details MUST be listed under comments)		

TODAY'S DATE	JOB/PROJECT INFORMATION
INITIALS	

HIGH TEMP		
LOW TEMP		

SCHEDULE INFORMATION	INSPECTION
Date of projected completion: / /	
Is project on schedule: ☐ Yes ☐ No	
Days behind:	

JOBSITE NOTES

INJURIES

Injuries on the job:	☐ Yes	☐ No
If yes, was OSHA notified:	☐ Yes	☐ No
Type of injury:	☐ First Aid	☐ Hospital

(*Details MUST be listed under comments*)

SAFETY

Toolbox topic:	Signage posted: ☐ Yes ☐ No
Everyone wearing PPE: ☐ Yes ☐ No	Checklist Complete: ☐ Yes ☐ No

TODAY'S DATE	JOB/PROJECT INFORMATION
INITIALS	

HIGH TEMP LOW TEMP									

SCHEDULE INFORMATION	INSPECTION
Date of projected completion: / /	
Is project on schedule: ☐ Yes ☐ No	
Days behind:	

JOBSITE NOTES

INJURIES		SAFETY	
Injuries on the job: ☐ Yes ☐ No		Toolbox topic:	Signage posted: ☐ Yes ☐ No
If yes, was OSHA notified: ☐ Yes ☐ No		Everyone wearing PPE: ☐ Yes ☐ No	Checklist Complete: ☐ Yes ☐ No
Type of injury: ☐ First Aid ☐ Hospital *(Details MUST be listed under comments)*			

TODAY'S DATE	JOB/PROJECT INFORMATION
INITIALS	

HIGH TEMP									
LOW TEMP									

SCHEDULE INFORMATION	INSPECTION
Date of projected completion: / /	
Is project on schedule: ☐ Yes ☐ No	
Days behind:	

JOBSITE NOTES

(blank lines)

INJURIES

Injuries on the job: ☐ Yes   ☐ No

If yes, was OSHA notified: ☐ Yes   ☐ No

Type of injury: ☐ First Aid   ☐ Hospital
(*Details MUST be listed under comments*)

SAFETY

Toolbox topic:	Signage posted: ☐ Yes ☐ No
Everyone wearing PPE: ☐ Yes ☐ No	Checklist Complete: ☐ Yes ☐ No

TODAY'S DATE	JOB/PROJECT INFORMATION
INITIALS	

HIGH TEMP									
LOW TEMP									

SCHEDULE INFORMATION	INSPECTION
Date of projected completion: / /	
Is project on schedule: ☐ Yes ☐ No	
Days behind:	

JOBSITE NOTES

INJURIES		SAFETY	
Injuries on the job: ☐ Yes ☐ No		Toolbox topic:	Signage posted: ☐ Yes ☐ No
If yes, was OSHA notified: ☐ Yes ☐ No			
Type of injury: ☐ First Aid ☐ Hospital		Everyone wearing PPE: ☐ Yes ☐ No	Checklist Complete: ☐ Yes ☐ No
(*Details MUST be listed under comments*)			

TODAY'S DATE	JOB/PROJECT INFORMATION
INITIALS	

HIGH TEMP									
LOW TEMP									

SCHEDULE INFORMATION	INSPECTION
Date of projected completion: / /	
Is project on schedule: ☐ Yes ☐ No	
Days behind:	

JOBSITE NOTES

INJURIES

Injuries on the job: ☐ Yes ☐ No	
If yes, was OSHA notified: ☐ Yes ☐ No	
Type of injury: ☐ First Aid ☐ Hospital	

(*Details MUST be listed under comments*)

SAFETY

Toolbox topic:	Signage posted: ☐ Yes ☐ No
Everyone wearing PPE: ☐ Yes ☐ No	Checklist Complete: ☐ Yes ☐ No

TODAY'S DATE	JOB/PROJECT INFORMATION
INITIALS	

			HIGH TEMP / LOW TEMP weather icons							

| **HIGH TEMP** | | | | | | | | | | |
| **LOW TEMP** | | | | | | | | | | |

SCHEDULE INFORMATION	INSPECTION
Date of projected completion: / /	
Is project on schedule: ☐ Yes ☐ No	
Days behind:	

JOBSITE NOTES

INJURIES		SAFETY	
Injuries on the job: ☐ Yes ☐ No		Toolbox topic:	Signage posted: ☐ Yes ☐ No
If yes, was OSHA notified: ☐ Yes ☐ No			
Type of injury: ☐ First Aid ☐ Hospital		Everyone wearing PPE: ☐ Yes ☐ No	Checklist Complete: ☐ Yes ☐ No
(Details MUST be listed under comments)			

TODAY'S DATE	JOB/PROJECT INFORMATION
INITIALS	

HIGH TEMP LOW TEMP								

SCHEDULE INFORMATION	INSPECTION
Date of projected completion: / /	
Is project on schedule: ☐ Yes ☐ No	
Days behind:	

JOBSITE NOTES

INJURIES

Injuries on the job: ☐ Yes ☐ No

If yes, was OSHA notified: ☐ Yes ☐ No

Type of injury: ☐ First Aid ☐ Hospital
(*Details MUST be listed under comments*)

SAFETY

Toolbox topic:	Signage posted: ☐ Yes ☐ No
Everyone wearing PPE: ☐ Yes ☐ No	Checklist Complete: ☐ Yes ☐ No

TODAY'S DATE	JOB/PROJECT INFORMATION
INITIALS	

HIGH TEMP									
LOW TEMP									

SCHEDULE INFORMATION	INSPECTION
Date of projected completion: / /	
Is project on schedule: ☐ Yes ☐ No	
Days behind:	

JOBSITE NOTES

INJURIES	SAFETY	
Injuries on the job: ☐ Yes ☐ No	Toolbox topic:	Signage posted: ☐ Yes ☐ No
If yes, was OSHA notified: ☐ Yes ☐ No		
Type of injury: ☐ First Aid ☐ Hospital *(Details MUST be listed under comments)*	Everyone wearing PPE: ☐ Yes ☐ No	Checklist Complete: ☐ Yes ☐ No

TODAY'S DATE	JOB/PROJECT INFORMATION
INITIALS	

HIGH TEMP									
LOW TEMP									

SCHEDULE INFORMATION	INSPECTION
Date of projected completion: / /	
Is project on schedule: ☐ Yes ☐ No	
Days behind:	

JOBSITE NOTES

_____

_____

_____

_____

_____

_____

_____

_____

_____

_____

_____

_____

_____

_____

_____

_____

_____

_____

INJURIES	SAFETY	
Injuries on the job: ☐ Yes ☐ No	Toolbox topic:	Signage posted: ☐ Yes ☐ No
If yes, was OSHA notified: ☐ Yes ☐ No		
Type of injury: ☐ First Aid ☐ Hospital	Everyone wearing PPE: ☐ Yes ☐ No	Checklist Complete: ☐ Yes ☐ No
(*Details MUST be listed under comments*)		

TODAY'S DATE	JOB/PROJECT INFORMATION
INITIALS	

HIGH TEMP									
LOW TEMP									

SCHEDULE INFORMATION	INSPECTION
Date of projected completion: / /	
Is project on schedule: ☐ Yes ☐ No	
Days behind:	

JOBSITE NOTES

INJURIES
Injuries on the job: ☐ Yes   ☐ No
If yes, was OSHA notified: ☐ Yes   ☐ No
Type of injury: ☐ First Aid   ☐ Hospital
(Details MUST be listed under comments)

SAFETY	
Toolbox topic:	Signage posted: ☐ Yes ☐ No
Everyone wearing PPE: ☐ Yes ☐ No	Checklist Complete: ☐ Yes ☐ No

TODAY'S DATE	JOB/PROJECT INFORMATION
INITIALS	

HIGH TEMP		
LOW TEMP		

SCHEDULE INFORMATION	INSPECTION
Date of projected completion: / /	
Is project on schedule: ☐ Yes ☐ No	
Days behind:	

JOBSITE NOTES

INJURIES

Injuries on the job: ☐ Yes ☐ No	
If yes, was OSHA notified: ☐ Yes ☐ No	
Type of injury: ☐ First Aid ☐ Hospital	
(*Details MUST be listed under comments*)	

SAFETY

Toolbox topic:	Signage posted: ☐ Yes ☐ No
Everyone wearing PPE: ☐ Yes ☐ No	Checklist Complete: ☐ Yes ☐ No

TODAY'S DATE	JOB/PROJECT INFORMATION
INITIALS	

HIGH TEMP		
LOW TEMP		

SCHEDULE INFORMATION	INSPECTION
Date of projected completion: / /	
Is project on schedule: ☐ Yes ☐ No	
Days behind:	

JOBSITE NOTES

INJURIES	SAFETY	
Injuries on the job: ☐ Yes ☐ No	Toolbox topic:	Signage posted: ☐ Yes ☐ No
If yes, was OSHA notified: ☐ Yes ☐ No		
Type of injury: ☐ First Aid ☐ Hospital *(Details MUST be listed under comments)*	Everyone wearing PPE: ☐ Yes ☐ No	Checklist Complete: ☐ Yes ☐ No

TODAY'S DATE	JOB/PROJECT INFORMATION
INITIALS	

HIGH TEMP LOW TEMP								

SCHEDULE INFORMATION

SCHEDULE INFORMATION	INSPECTION
Date of projected completion: / /	
Is project on schedule: ☐ Yes ☐ No	
Days behind:	

JOBSITE NOTES

INJURIES

INJURIES
Injuries on the job: ☐ Yes ☐ No
If yes, was OSHA notified: ☐ Yes ☐ No
Type of injury: ☐ First Aid ☐ Hospital (*Details MUST be listed under comments*)

SAFETY

SAFETY	
Toolbox topic:	Signage posted: ☐ Yes ☐ No
Everyone wearing PPE: ☐ Yes ☐ No	Checklist Complete: ☐ Yes ☐ No

TODAY'S DATE	JOB/PROJECT INFORMATION
INITIALS	

HIGH TEMP LOW TEMP									

SCHEDULE INFORMATION	INSPECTION
Date of projected completion: / /	
Is project on schedule: ☐ Yes ☐ No	
Days behind:	

JOBSITE NOTES

INJURIES		SAFETY	
Injuries on the job: ☐ Yes ☐ No		Toolbox topic:	Signage posted: ☐ Yes ☐ No
If yes, was OSHA notified: ☐ Yes ☐ No			
Type of injury: ☐ First Aid ☐ Hospital		Everyone wearing PPE: ☐ Yes ☐ No	Checklist Complete: ☐ Yes ☐ No
(Details MUST be listed under comments)			

TODAY'S DATE	JOB/PROJECT INFORMATION
INITIALS	

HIGH TEMP LOW TEMP								

SCHEDULE INFORMATION	INSPECTION
Date of projected completion: / /	
Is project on schedule: ☐ Yes ☐ No	
Days behind:	

JOBSITE NOTES

INJURIES		SAFETY	
Injuries on the job: ☐ Yes ☐ No		Toolbox topic:	Signage posted: ☐ Yes ☐ No
If yes, was OSHA notified: ☐ Yes ☐ No			
Type of injury: ☐ First Aid ☐ Hospital (*Details MUST be listed under comments*)		Everyone wearing PPE: ☐ Yes ☐ No	Checklist Complete: ☐ Yes ☐ No

TODAY'S DATE	JOB/PROJECT INFORMATION
INITIALS	

HIGH TEMP LOW TEMP								

SCHEDULE INFORMATION	INSPECTION
Date of projected completion: / /	
Is project on schedule: ☐ Yes ☐ No	
Days behind:	

JOBSITE NOTES

INJURIES	
Injuries on the job: ☐ Yes ☐ No	
If yes, was OSHA notified: ☐ Yes ☐ No	
Type of injury: ☐ First Aid ☐ Hospital	
(Details MUST be listed under comments)	

SAFETY	
Toolbox topic:	Signage posted: ☐ Yes ☐ No
Everyone wearing PPE: ☐ Yes ☐ No	Checklist Complete: ☐ Yes ☐ No

TODAY'S DATE	JOB/PROJECT INFORMATION
INITIALS	

HIGH TEMP									
LOW TEMP									

SCHEDULE INFORMATION	INSPECTION
Date of projected completion: / /	
Is project on schedule: ☐ Yes ☐ No	
Days behind:	

JOBSITE NOTES

(blank lined notes area)

INJURIES

Injuries on the job: ☐ Yes ☐ No	
If yes, was OSHA notified: ☐ Yes ☐ No	
Type of injury: ☐ First Aid ☐ Hospital	
(Details MUST be listed under comments)	

SAFETY

Toolbox topic:	Signage posted: ☐ Yes ☐ No
Everyone wearing PPE: ☐ Yes ☐ No	Checklist Complete: ☐ Yes ☐ No

TODAY'S DATE	JOB/PROJECT INFORMATION
INITIALS	

HIGH TEMP LOW TEMP									

SCHEDULE INFORMATION	INSPECTION
Date of projected completion: / /	
Is project on schedule: ☐ Yes ☐ No	
Days behind:	

JOBSITE NOTES

INJURIES		SAFETY	
Injuries on the job: ☐ Yes ☐ No		Toolbox topic:	Signage posted: ☐ Yes ☐ No
If yes, was OSHA notified: ☐ Yes ☐ No			
Type of injury: ☐ First Aid ☐ Hospital		Everyone wearing PPE: ☐ Yes ☐ No	Checklist Complete: ☐ Yes ☐ No
(Details MUST be listed under comments)			

TODAY'S DATE	JOB/PROJECT INFORMATION
INITIALS	

HIGH TEMP									
LOW TEMP									

SCHEDULE INFORMATION	INSPECTION
Date of projected completion: / /	
Is project on schedule: ☐ Yes ☐ No	
Days behind:	

JOBSITE NOTES

INJURIES		SAFETY	
Injuries on the job: ☐ Yes ☐ No		Toolbox topic:	Signage posted: ☐ Yes ☐ No
If yes, was OSHA notified: ☐ Yes ☐ No			
Type of injury: ☐ First Aid ☐ Hospital (*Details MUST be listed under comments*)		Everyone wearing PPE: ☐ Yes ☐ No	Checklist Complete: ☐ Yes ☐ No

TODAY'S DATE	JOB/PROJECT INFORMATION
INITIALS	

HIGH TEMP										
LOW TEMP										

SCHEDULE INFORMATION	INSPECTION
Date of projected completion: / /	
Is project on schedule: ☐ Yes ☐ No	
Days behind:	

JOBSITE NOTES

INJURIES		SAFETY	
Injuries on the job: ☐ Yes ☐ No		Toolbox topic:	Signage posted: ☐ Yes ☐ No
If yes, was OSHA notified: ☐ Yes ☐ No			
Type of injury: ☐ First Aid ☐ Hospital *(Details MUST be listed under comments)*		Everyone wearing PPE: ☐ Yes ☐ No	Checklist Complete: ☐ Yes ☐ No

TODAY'S DATE	JOB/PROJECT INFORMATION
INITIALS	

HIGH TEMP		
LOW TEMP		

SCHEDULE INFORMATION	INSPECTION
Date of projected completion: / /	
Is project on schedule: ☐ Yes ☐ No	
Days behind:	

JOBSITE NOTES

INJURIES

Injuries on the job: ☐ Yes ☐ No

If yes, was OSHA notified: ☐ Yes ☐ No

Type of injury: ☐ First Aid ☐ Hospital
(*Details MUST be listed under comments*)

SAFETY

Toolbox topic:	Signage posted: ☐ Yes ☐ No
Everyone wearing PPE: ☐ Yes ☐ No	Checklist Complete: ☐ Yes ☐ No

TODAY'S DATE	JOB/PROJECT INFORMATION
INITIALS	

HIGH TEMP									
LOW TEMP									

SCHEDULE INFORMATION	INSPECTION
Date of projected completion: / /	
Is project on schedule: ☐ Yes ☐ No	
Days behind:	

JOBSITE NOTES

INJURIES

Injuries on the job: ☐ Yes   ☐ No

If yes, was OSHA notified: ☐ Yes   ☐ No

Type of injury: ☐ First Aid   ☐ Hospital
(*Details MUST be listed under comments*)

SAFETY

Toolbox topic:	Signage posted: ☐ Yes ☐ No
Everyone wearing PPE: ☐ Yes ☐ No	Checklist Complete: ☐ Yes ☐ No

TODAY'S DATE		JOB/PROJECT INFORMATION
INITIALS		

HIGH TEMP									
LOW TEMP									

SCHEDULE INFORMATION	INSPECTION
Date of projected completion: / /	
Is project on schedule: ☐ Yes ☐ No	
Days behind:	

JOBSITE NOTES

INJURIES

Injuries on the job: ☐ Yes   ☐ No

If yes, was OSHA notified: ☐ Yes   ☐ No

Type of injury: ☐ First Aid   ☐ Hospital
(*Details MUST be listed under comments*)

SAFETY

Toolbox topic:	Signage posted: ☐ Yes ☐ No
Everyone wearing PPE: ☐ Yes ☐ No	Checklist Complete: ☐ Yes ☐ No

TODAY'S DATE	JOB/PROJECT INFORMATION
INITIALS	

HIGH TEMP									
LOW TEMP									

SCHEDULE INFORMATION	INSPECTION
Date of projected completion: / /	
Is project on schedule: ☐ Yes ☐ No	
Days behind:	

JOBSITE NOTES

INJURIES		SAFETY	
Injuries on the job: ☐ Yes ☐ No		Toolbox topic:	Signage posted: ☐ Yes ☐ No
If yes, was OSHA notified: ☐ Yes ☐ No			
Type of injury: ☐ First Aid ☐ Hospital		Everyone wearing PPE: ☐ Yes ☐ No	Checklist Complete: ☐ Yes ☐ No
(*Details MUST be listed under comments*)			

TODAY'S DATE	JOB/PROJECT INFORMATION
INITIALS	

HIGH TEMP LOW TEMP									

SCHEDULE INFORMATION	INSPECTION
Date of projected completion: / /	
Is project on schedule: ☐ Yes ☐ No	
Days behind:	

JOBSITE NOTES

INJURIES		SAFETY	
Injuries on the job: ☐ Yes ☐ No		Toolbox topic:	Signage posted: ☐ Yes ☐ No
If yes, was OSHA notified: ☐ Yes ☐ No			
Type of injury: ☐ First Aid ☐ Hospital *(Details MUST be listed under comments)*		Everyone wearing PPE: ☐ Yes ☐ No	Checklist Complete: ☐ Yes ☐ No

TODAY'S DATE	JOB/PROJECT INFORMATION
INITIALS	

HIGH TEMP **LOW TEMP**								

SCHEDULE INFORMATION	INSPECTION
Date of projected completion: / /	
Is project on schedule: ☐ Yes ☐ No	
Days behind:	

JOBSITE NOTES

INJURIES		SAFETY	
Injuries on the job: ☐ Yes ☐ No		Toolbox topic:	Signage posted: ☐ Yes ☐ No
If yes, was OSHA notified: ☐ Yes ☐ No			
Type of injury: ☐ First Aid ☐ Hospital *(Details MUST be listed under comments)*		Everyone wearing PPE: ☐ Yes ☐ No	Checklist Complete: ☐ Yes ☐ No

TODAY'S DATE	JOB/PROJECT INFORMATION
INITIALS	

HIGH TEMP									
LOW TEMP									

SCHEDULE INFORMATION	INSPECTION
Date of projected completion: / /	
Is project on schedule: ☐ Yes ☐ No	
Days behind:	

JOBSITE NOTES

INJURIES		SAFETY	
Injuries on the job: ☐ Yes ☐ No		Toolbox topic:	Signage posted: ☐ Yes ☐ No
If yes, was OSHA notified: ☐ Yes ☐ No			
Type of injury: ☐ First Aid ☐ Hospital		Everyone wearing PPE: ☐ Yes ☐ No	Checklist Complete: ☐ Yes ☐ No
(*Details MUST be listed under comments*)			

TODAY'S DATE	JOB/PROJECT INFORMATION
INITIALS	

HIGH TEMP LOW TEMP								

SCHEDULE INFORMATION	INSPECTION
Date of projected completion: / /	
Is project on schedule: ☐ Yes ☐ No	
Days behind:	

JOBSITE NOTES

INJURIES		SAFETY	
Injuries on the job: ☐ Yes ☐ No		Toolbox topic:	Signage posted: ☐ Yes ☐ No
If yes, was OSHA notified: ☐ Yes ☐ No		Everyone wearing PPE: ☐ Yes ☐ No	Checklist Complete: ☐ Yes ☐ No
Type of injury: ☐ First Aid ☐ Hospital *(Details MUST be listed under comments)*			

TODAY'S DATE	JOB/PROJECT INFORMATION
INITIALS	

HIGH TEMP	
LOW TEMP	

SCHEDULE INFORMATION	INSPECTION
Date of projected completion: / /	
Is project on schedule: ☐ Yes ☐ No	
Days behind:	

JOBSITE NOTES

INJURIES

Injuries on the job: ☐ Yes   ☐ No

If yes, was OSHA notified: ☐ Yes   ☐ No

Type of injury: ☐ First Aid   ☐ Hospital
(*Details MUST be listed under comments*)

SAFETY

Toolbox topic:	Signage posted: ☐ Yes ☐ No
Everyone wearing PPE: ☐ Yes ☐ No	Checklist Complete: ☐ Yes ☐ No

TODAY'S DATE	JOB/PROJECT INFORMATION
INITIALS	

HIGH TEMP LOW TEMP									

SCHEDULE INFORMATION	INSPECTION
Date of projected completion: / /	
Is project on schedule: ☐ Yes ☐ No	
Days behind:	

JOBSITE NOTES

INJURIES		SAFETY	
Injuries on the job: ☐ Yes ☐ No		Toolbox topic:	Signage posted: ☐ Yes ☐ No
If yes, was OSHA notified: ☐ Yes ☐ No			
Type of injury: ☐ First Aid ☐ Hospital		Everyone wearing PPE: ☐ Yes ☐ No	Checklist Complete: ☐ Yes ☐ No
(Details MUST be listed under comments)			

MAIN LOG

TODAY'S DATE	JOB/PROJECT INFORMATION
INITIALS	

HIGH TEMP		
LOW TEMP		

SCHEDULE INFORMATION	INSPECTION
Date of projected completion: / /	
Is project on schedule: ☐ Yes ☐ No	
Days behind:	

JOBSITE NOTES

INJURIES

Injuries on the job: ☐ Yes   ☐ No

If yes, was OSHA notified: ☐ Yes   ☐ No

Type of injury: ☐ First Aid   ☐ Hospital
(*Details MUST be listed under comments*)

SAFETY

Toolbox topic:	Signage posted: ☐ Yes ☐ No
Everyone wearing PPE: ☐ Yes ☐ No	Checklist Complete: ☐ Yes ☐ No

TODAY'S DATE	JOB/PROJECT INFORMATION
INITIALS	

HIGH TEMP									
LOW TEMP									

SCHEDULE INFORMATION	INSPECTION
Date of projected completion: / /	
Is project on schedule: ☐ Yes ☐ No	
Days behind:	

JOBSITE NOTES

INJURIES		SAFETY	
Injuries on the job: ☐ Yes ☐ No		Toolbox topic:	Signage posted: ☐ Yes ☐ No
If yes, was OSHA notified: ☐ Yes ☐ No			
Type of injury: ☐ First Aid ☐ Hospital (*Details MUST be listed under comments*)		Everyone wearing PPE: ☐ Yes ☐ No	Checklist Complete: ☐ Yes ☐ No

TODAY'S DATE		JOB/PROJECT INFORMATION
INITIALS		

HIGH TEMP LOW TEMP								

SCHEDULE INFORMATION / INSPECTION

Date of projected completion: /   /

Is project on schedule: ☐ Yes ☐ No

Days behind:

JOBSITE NOTES

INJURIES

Injuries on the job: ☐ Yes ☐ No

If yes, was OSHA notified: ☐ Yes ☐ No

Type of injury: ☐ First Aid ☐ Hospital
(*Details MUST be listed under comments*)

SAFETY

Toolbox topic:

Signage posted: ☐ Yes ☐ No

Everyone wearing PPE: ☐ Yes ☐ No

Checklist Complete: ☐ Yes ☐ No

TODAY'S DATE	JOB/PROJECT INFORMATION
INITIALS	

HIGH TEMP LOW TEMP									

SCHEDULE INFORMATION	INSPECTION
Date of projected completion: / /	
Is project on schedule: ☐ Yes ☐ No	
Days behind:	

JOBSITE NOTES

INJURIES	SAFETY	
Injuries on the job: ☐ Yes ☐ No	Toolbox topic:	Signage posted: ☐ Yes ☐ No
If yes, was OSHA notified: ☐ Yes ☐ No		
Type of injury: ☐ First Aid ☐ Hospital	Everyone wearing PPE: ☐ Yes ☐ No	Checklist Complete: ☐ Yes ☐ No
(*Details MUST be listed under comments*)		

TODAY'S DATE	JOB/PROJECT INFORMATION
INITIALS	

HIGH TEMP									
LOW TEMP									

SCHEDULE INFORMATION	INSPECTION
Date of projected completion: / /	
Is project on schedule: ☐ Yes ☐ No	
Days behind:	

JOBSITE NOTES

INJURIES		SAFETY	
Injuries on the job: ☐ Yes ☐ No		Toolbox topic:	Signage posted: ☐ Yes ☐ No
If yes, was OSHA notified: ☐ Yes ☐ No			
Type of injury: ☐ First Aid ☐ Hospital		Everyone wearing PPE: ☐ Yes ☐ No	Checklist Complete: ☐ Yes ☐ No
(Details MUST be listed under comments)			

TODAY'S DATE	JOB/PROJECT INFORMATION
INITIALS	

HIGH TEMP									
LOW TEMP									

SCHEDULE INFORMATION	INSPECTION
Date of projected completion: / /	
Is project on schedule: ☐ Yes ☐ No	
Days behind:	

JOBSITE NOTES

INJURIES		SAFETY	
Injuries on the job: ☐ Yes ☐ No		Toolbox topic:	Signage posted: ☐ Yes ☐ No
If yes, was OSHA notified: ☐ Yes ☐ No		Everyone wearing PPE: ☐ Yes ☐ No	Checklist Complete: ☐ Yes ☐ No
Type of injury: ☐ First Aid ☐ Hospital *(Details MUST be listed under comments)*			

TODAY'S DATE	JOB/PROJECT INFORMATION
INITIALS	

HIGH TEMP									
LOW TEMP									

SCHEDULE INFORMATION	INSPECTION
Date of projected completion: / /	
Is project on schedule: ☐ Yes ☐ No	
Days behind:	

JOBSITE NOTES

INJURIES	
Injuries on the job: ☐ Yes ☐ No	
If yes, was OSHA notified: ☐ Yes ☐ No	
Type of injury: ☐ First Aid ☐ Hospital	
(*Details MUST be listed under comments*)	

SAFETY	
Toolbox topic:	Signage posted: ☐ Yes ☐ No
Everyone wearing PPE: ☐ Yes ☐ No	Checklist Complete: ☐ Yes ☐ No

TODAY'S DATE	JOB/PROJECT INFORMATION
INITIALS	

		HIGH TEMP							
HIGH TEMP									
LOW TEMP									

SCHEDULE INFORMATION	INSPECTION
Date of projected completion: / /	
Is project on schedule: ☐ Yes ☐ No	
Days behind:	

JOBSITE NOTES

(blank lined notes area)

INJURIES

Injuries on the job: ☐ Yes   ☐ No

If yes, was OSHA notified: ☐ Yes   ☐ No

Type of injury: ☐ First Aid   ☐ Hospital
(Details MUST be listed under comments)

SAFETY

Toolbox topic:	Signage posted: ☐ Yes ☐ No
Everyone wearing PPE: ☐ Yes ☐ No	Checklist Complete: ☐ Yes ☐ No

TODAY'S DATE	JOB/PROJECT INFORMATION
INITIALS	

HIGH TEMP									
LOW TEMP									

SCHEDULE INFORMATION	INSPECTION
Date of projected completion: / /	
Is project on schedule: ☐ Yes ☐ No	
Days behind:	

JOBSITE NOTES

_____

_____

_____

_____

_____

_____

_____

_____

_____

_____

_____

_____

_____

_____

_____

_____

INJURIES		SAFETY	
Injuries on the job: ☐ Yes ☐ No		Toolbox topic:	Signage posted: ☐ Yes ☐ No
If yes, was OSHA notified: ☐ Yes ☐ No			
Type of injury: ☐ First Aid ☐ Hospital _(Details MUST be listed under comments)_		Everyone wearing PPE: ☐ Yes ☐ No	Checklist Complete: ☐ Yes ☐ No

TODAY'S DATE	JOB/PROJECT INFORMATION
INITIALS	

HIGH TEMP									
LOW TEMP									

SCHEDULE INFORMATION	INSPECTION
Date of projected completion: / /	
Is project on schedule: ☐ Yes ☐ No	
Days behind:	

JOBSITE NOTES

INJURIES	
Injuries on the job: ☐ Yes ☐ No	
If yes, was OSHA notified: ☐ Yes ☐ No	
Type of injury: ☐ First Aid ☐ Hospital	
(Details MUST be listed under comments)	

SAFETY	
Toolbox topic:	Signage posted: ☐ Yes ☐ No
Everyone wearing PPE: ☐ Yes ☐ No	Checklist Complete: ☐ Yes ☐ No

TODAY'S DATE	JOB/PROJECT INFORMATION
INITIALS	

HIGH TEMP		
LOW TEMP		

SCHEDULE INFORMATION	INSPECTION
Date of projected completion: / /	
Is project on schedule: ☐ Yes ☐ No	
Days behind:	

JOBSITE NOTES

INJURIES

Injuries on the job: ☐ Yes ☐ No	
If yes, was OSHA notified: ☐ Yes ☐ No	
Type of injury: ☐ First Aid ☐ Hospital	
(Details MUST be listed under comments)	

SAFETY

Toolbox topic:	Signage posted: ☐ Yes ☐ No
Everyone wearing PPE: ☐ Yes ☐ No	Checklist Complete: ☐ Yes ☐ No

TODAY'S DATE	JOB/PROJECT INFORMATION
INITIALS	

HIGH TEMP									
LOW TEMP									

SCHEDULE INFORMATION | INSPECTION

SCHEDULE INFORMATION	INSPECTION
Date of projected completion: / /	
Is project on schedule: ☐ Yes ☐ No	
Days behind:	

JOBSITE NOTES

JOBSITE NOTES

INJURIES

INJURIES
Injuries on the job: ☐ Yes   ☐ No
If yes, was OSHA notified: ☐ Yes   ☐ No
Type of injury: ☐ First Aid   ☐ Hospital
(*Details MUST be listed under comments*)

SAFETY

SAFETY	
Toolbox topic:	Signage posted: ☐ Yes ☐ No
Everyone wearing PPE: ☐ Yes ☐ No	Checklist Complete: ☐ Yes ☐ No

TODAY'S DATE	JOB/PROJECT INFORMATION
INITIALS	

HIGH TEMP LOW TEMP								

SCHEDULE INFORMATION	INSPECTION
Date of projected completion: / /	
Is project on schedule: ☐ Yes ☐ No	
Days behind:	

JOBSITE NOTES

INJURIES		SAFETY	
Injuries on the job: ☐ Yes ☐ No		Toolbox topic:	Signage posted: ☐ Yes ☐ No
If yes, was OSHA notified: ☐ Yes ☐ No			
Type of injury: ☐ First Aid ☐ Hospital (*Details MUST be listed under comments*)		Everyone wearing PPE: ☐ Yes ☐ No	Checklist Complete: ☐ Yes ☐ No

TODAY'S DATE	JOB/PROJECT INFORMATION
INITIALS	

HIGH TEMP									
LOW TEMP									

SCHEDULE INFORMATION	INSPECTION
Date of projected completion: / /	
Is project on schedule: ☐ Yes ☐ No	
Days behind:	

JOBSITE NOTES

INJURIES		SAFETY	
Injuries on the job: ☐ Yes ☐ No		Toolbox topic:	Signage posted: ☐ Yes ☐ No
If yes, was OSHA notified: ☐ Yes ☐ No			
Type of injury: ☐ First Aid ☐ Hospital		Everyone wearing PPE: ☐ Yes ☐ No	Checklist Complete: ☐ Yes ☐ No
(*Details MUST be listed under comments*)			

TODAY'S DATE	JOB/PROJECT INFORMATION
INITIALS	

HIGH TEMP LOW TEMP									

MAIN LOG

SCHEDULE INFORMATION	INSPECTION
Date of projected completion: / /	
Is project on schedule: ☐ Yes ☐ No	
Days behind:	

JOBSITE NOTES

INJURIES

Injuries on the job: ☐ Yes ☐ No

If yes, was OSHA notified: ☐ Yes ☐ No

Type of injury: ☐ First Aid ☐ Hospital

(*Details MUST be listed under comments*)

SAFETY

Toolbox topic:	Signage posted: ☐ Yes ☐ No
Everyone wearing PPE: ☐ Yes ☐ No	Checklist Complete: ☐ Yes ☐ No

TODAY'S DATE	JOB/PROJECT INFORMATION
INITIALS	

HIGH TEMP LOW TEMP									

SCHEDULE INFORMATION	INSPECTION
Date of projected completion: / /	
Is project on schedule: ☐ Yes ☐ No	
Days behind:	

JOBSITE NOTES

(blank lined notes area)

INJURIES	
Injuries on the job: ☐ Yes ☐ No	
If yes, was OSHA notified: ☐ Yes ☐ No	
Type of injury: ☐ First Aid ☐ Hospital	
(Details MUST be listed under comments)	

SAFETY	
Toolbox topic:	Signage posted: ☐ Yes ☐ No
Everyone wearing PPE: ☐ Yes ☐ No	Checklist Complete: ☐ Yes ☐ No

TODAY'S DATE	JOB/PROJECT INFORMATION
INITIALS	

HIGH TEMP		
LOW TEMP		

SCHEDULE INFORMATION	INSPECTION
Date of projected completion: / /	
Is project on schedule: ☐ Yes ☐ No	
Days behind:	

JOBSITE NOTES

INJURIES

Injuries on the job: ☐ Yes   ☐ No

If yes, was OSHA notified: ☐ Yes   ☐ No

Type of injury: ☐ First Aid   ☐ Hospital

(*Details MUST be listed under comments*)

SAFETY

Toolbox topic:	Signage posted: ☐ Yes ☐ No
Everyone wearing PPE: ☐ Yes ☐ No	Checklist Complete: ☐ Yes ☐ No

TODAY'S DATE	JOB/PROJECT INFORMATION
INITIALS	

HIGH TEMP LOW TEMP									

SCHEDULE INFORMATION	INSPECTION
Date of projected completion: / /	
Is project on schedule: ☐ Yes ☐ No	
Days behind:	

JOBSITE NOTES

INJURIES		SAFETY	
Injuries on the job: ☐ Yes ☐ No		Toolbox topic:	Signage posted: ☐ Yes ☐ No
If yes, was OSHA notified: ☐ Yes ☐ No		Everyone wearing PPE: ☐ Yes ☐ No	Checklist Complete: ☐ Yes ☐ No
Type of injury: ☐ First Aid ☐ Hospital *(Details MUST be listed under comments)*			

TODAY'S DATE	JOB/PROJECT INFORMATION
INITIALS	

HIGH TEMP		
LOW TEMP		

SCHEDULE INFORMATION	INSPECTION
Date of projected completion: / /	
Is project on schedule: ☐ Yes ☐ No	
Days behind:	

JOBSITE NOTES

INJURIES

Injuries on the job: ☐ Yes   ☐ No

If yes, was OSHA notified: ☐ Yes   ☐ No

Type of injury: ☐ First Aid   ☐ Hospital
(*Details MUST be listed under comments*)

SAFETY

Toolbox topic:	Signage posted: ☐ Yes ☐ No
Everyone wearing PPE: ☐ Yes ☐ No	Checklist Complete: ☐ Yes ☐ No

TODAY'S DATE	JOB/PROJECT INFORMATION
INITIALS	

HIGH TEMP LOW TEMP									

SCHEDULE INFORMATION	INSPECTION
Date of projected completion: / /	
Is project on schedule: ☐ Yes ☐ No	
Days behind:	

JOBSITE NOTES

INJURIES		SAFETY	
Injuries on the job: ☐ Yes ☐ No		Toolbox topic:	Signage posted: ☐ Yes ☐ No
If yes, was OSHA notified: ☐ Yes ☐ No			
Type of injury: ☐ First Aid ☐ Hospital		Everyone wearing PPE: ☐ Yes ☐ No	Checklist Complete: ☐ Yes ☐ No
(*Details MUST be listed under comments*)			

TODAY'S DATE	JOB/PROJECT INFORMATION
INITIALS	

HIGH TEMP									
LOW TEMP									

SCHEDULE INFORMATION	INSPECTION
Date of projected completion: / /	
Is project on schedule: ☐ Yes ☐ No	
Days behind:	

JOBSITE NOTES

_____
_____
_____
_____
_____
_____
_____
_____
_____
_____
_____
_____
_____
_____
_____
_____
_____
_____

INJURIES	
Injuries on the job: ☐ Yes ☐ No	
If yes, was OSHA notified: ☐ Yes ☐ No	
Type of injury: ☐ First Aid ☐ Hospital	
(*Details MUST be listed under comments*)	

SAFETY	
Toolbox topic:	Signage posted: ☐ Yes ☐ No
Everyone wearing PPE: ☐ Yes ☐ No	Checklist Complete: ☐ Yes ☐ No

TODAY'S DATE	JOB/PROJECT INFORMATION
INITIALS	

HIGH TEMP										
LOW TEMP										

SCHEDULE INFORMATION	INSPECTION
Date of projected completion: / /	
Is project on schedule: ☐ Yes ☐ No	
Days behind:	

JOBSITE NOTES

INJURIES	SAFETY	
Injuries on the job: ☐ Yes ☐ No	Toolbox topic:	Signage posted: ☐ Yes ☐ No
If yes, was OSHA notified: ☐ Yes ☐ No		
Type of injury: ☐ First Aid ☐ Hospital	Everyone wearing PPE: ☐ Yes ☐ No	Checklist Complete: ☐ Yes ☐ No
(Details MUST be listed under comments)		

TODAY'S DATE	JOB/PROJECT INFORMATION
INITIALS	

HIGH TEMP LOW TEMP								

SCHEDULE INFORMATION	INSPECTION
Date of projected completion: / /	
Is project on schedule: ☐ Yes ☐ No	
Days behind:	

JOBSITE NOTES

INJURIES

Injuries on the job: ☐ Yes ☐ No

If yes, was OSHA notified: ☐ Yes ☐ No

Type of injury: ☐ First Aid ☐ Hospital
(*Details MUST be listed under comments*)

SAFETY

Toolbox topic:	Signage posted: ☐ Yes ☐ No
Everyone wearing PPE: ☐ Yes ☐ No	Checklist Complete: ☐ Yes ☐ No

TODAY'S DATE	JOB/PROJECT INFORMATION
INITIALS	

HIGH TEMP									
LOW TEMP									

SCHEDULE INFORMATION	INSPECTION
Date of projected completion: / /	
Is project on schedule: ☐ Yes ☐ No	
Days behind:	

JOBSITE NOTES

INJURIES		SAFETY	
Injuries on the job: ☐ Yes ☐ No		Toolbox topic:	Signage posted: ☐ Yes ☐ No
If yes, was OSHA notified: ☐ Yes ☐ No			
Type of injury: ☐ First Aid ☐ Hospital		Everyone wearing PPE: ☐ Yes ☐ No	Checklist Complete: ☐ Yes ☐ No
(Details MUST be listed under comments)			

TODAY'S DATE	JOB/PROJECT INFORMATION
INITIALS	

HIGH TEMP										
LOW TEMP										

SCHEDULE INFORMATION	INSPECTION
Date of projected completion: / /	
Is project on schedule: ☐ Yes ☐ No	
Days behind:	

JOBSITE NOTES

INJURIES

Injuries on the job: ☐ Yes ☐ No

If yes, was OSHA notified: ☐ Yes ☐ No

Type of injury: ☐ First Aid ☐ Hospital
(*Details MUST be listed under comments*)

SAFETY

Toolbox topic:	Signage posted: ☐ Yes ☐ No
Everyone wearing PPE: ☐ Yes ☐ No	Checklist Complete: ☐ Yes ☐ No

TODAY'S DATE	JOB/PROJECT INFORMATION
INITIALS	

HIGH TEMP									
LOW TEMP									

SCHEDULE INFORMATION	INSPECTION
Date of projected completion: / /	
Is project on schedule: ☐ Yes ☐ No	
Days behind:	

JOBSITE NOTES

_____
_____
_____
_____
_____
_____
_____
_____
_____
_____
_____
_____
_____
_____
_____
_____
_____
_____

INJURIES		SAFETY	
Injuries on the job: ☐ Yes ☐ No		Toolbox topic:	Signage posted: ☐ Yes ☐ No
If yes, was OSHA notified: ☐ Yes ☐ No			
Type of injury: ☐ First Aid ☐ Hospital _(Details MUST be listed under comments)_		Everyone wearing PPE: ☐ Yes ☐ No	Checklist Complete: ☐ Yes ☐ No

TODAY'S DATE	JOB/PROJECT INFORMATION
INITIALS	

HIGH TEMP										
LOW TEMP										

SCHEDULE INFORMATION	INSPECTION
Date of projected completion: / /	
Is project on schedule: ☐ Yes ☐ No	
Days behind:	

JOBSITE NOTES

INJURIES

Injuries on the job:	☐ Yes ☐ No
If yes, was OSHA notified:	☐ Yes ☐ No
Type of injury: ☐ First Aid ☐ Hospital	
(*Details MUST be listed under comments*)	

SAFETY

Toolbox topic:	Signage posted: ☐ Yes ☐ No
Everyone wearing PPE: ☐ Yes ☐ No	Checklist Complete: ☐ Yes ☐ No

TODAY'S DATE	JOB/PROJECT INFORMATION
INITIALS	

HIGH TEMP		
LOW TEMP		

SCHEDULE INFORMATION	INSPECTION
Date of projected completion: / /	
Is project on schedule: ☐ Yes ☐ No	
Days behind:	

JOBSITE NOTES

_____

_____

_____

_____

_____

_____

_____

_____

_____

_____

_____

_____

_____

_____

_____

_____

_____

_____

_____

INJURIES	SAFETY	
Injuries on the job: ☐ Yes ☐ No	Toolbox topic:	Signage posted: ☐ Yes ☐ No
If yes, was OSHA notified: ☐ Yes ☐ No		
Type of injury: ☐ First Aid ☐ Hospital	Everyone wearing PPE: ☐ Yes ☐ No	Checklist Complete: ☐ Yes ☐ No
(*Details MUST be listed under comments*)		

TODAY'S DATE	JOB/PROJECT INFORMATION

INITIALS

HIGH TEMP	
LOW TEMP	

SCHEDULE INFORMATION

Date of projected completion: /   /

Is project on schedule: ☐ Yes ☐ No

Days behind:

INSPECTION

JOBSITE NOTES

INJURIES

Injuries on the job: ☐ Yes   ☐ No

If yes, was OSHA notified: ☐ Yes   ☐ No

Type of injury: ☐ First Aid   ☐ Hospital
(*Details MUST be listed under comments*)

SAFETY

Toolbox topic:	Signage posted: ☐ Yes ☐ No
Everyone wearing PPE: ☐ Yes ☐ No	Checklist Complete: ☐ Yes ☐ No

TODAY'S DATE	JOB/PROJECT INFORMATION
INITIALS	

HIGH TEMP		
LOW TEMP		

SCHEDULE INFORMATION	INSPECTION
Date of projected completion: / /	
Is project on schedule: ☐ Yes ☐ No	
Days behind:	

JOBSITE NOTES

_____
_____
_____
_____
_____
_____
_____
_____
_____
_____
_____
_____
_____
_____
_____
_____
_____

INJURIES		SAFETY	
Injuries on the job: ☐ Yes ☐ No		Toolbox topic:	Signage posted: ☐ Yes ☐ No
If yes, was OSHA notified: ☐ Yes ☐ No			
Type of injury: ☐ First Aid ☐ Hospital _(Details MUST be listed under comments)_		Everyone wearing PPE: ☐ Yes ☐ No	Checklist Complete: ☐ Yes ☐ No

TODAY'S DATE	JOB/PROJECT INFORMATION
INITIALS	

HIGH TEMP **LOW TEMP**								

SCHEDULE INFORMATION	INSPECTION
Date of projected completion: / /	
Is project on schedule: ☐ Yes ☐ No	
Days behind:	

JOBSITE NOTES

INJURIES		SAFETY	
Injuries on the job: ☐ Yes ☐ No		Toolbox topic:	Signage posted: ☐ Yes ☐ No
If yes, was OSHA notified: ☐ Yes ☐ No			
Type of injury: ☐ First Aid ☐ Hospital		Everyone wearing PPE: ☐ Yes ☐ No	Checklist Complete: ☐ Yes ☐ No
(Details MUST be listed under comments)			

TODAY'S DATE	JOB/PROJECT INFORMATION
INITIALS	

HIGH TEMP		
LOW TEMP		

SCHEDULE INFORMATION	INSPECTION
Date of projected completion: / /	
Is project on schedule: ☐ Yes ☐ No	
Days behind:	

JOBSITE NOTES

INJURIES		
Injuries on the job: ☐ Yes ☐ No		
If yes, was OSHA notified: ☐ Yes ☐ No		
Type of injury: ☐ First Aid ☐ Hospital *(Details MUST be listed under comments)*		

SAFETY	
Toolbox topic:	Signage posted: ☐ Yes ☐ No
Everyone wearing PPE: ☐ Yes ☐ No	Checklist Complete: ☐ Yes ☐ No

TODAY'S DATE	JOB/PROJECT INFORMATION
INITIALS	

HIGH TEMP LOW TEMP									

SCHEDULE INFORMATION	INSPECTION
Date of projected completion: / /	
Is project on schedule: ☐ Yes ☐ No	
Days behind:	

JOBSITE NOTES

INJURIES
Injuries on the job: ☐ Yes ☐ No
If yes, was OSHA notified: ☐ Yes ☐ No
Type of injury: ☐ First Aid ☐ Hospital
(*Details MUST be listed under comments*)

SAFETY	
Toolbox topic:	Signage posted: ☐ Yes ☐ No
Everyone wearing PPE: ☐ Yes ☐ No	Checklist Complete: ☐ Yes ☐ No

TODAY'S DATE	JOB/PROJECT INFORMATION
INITIALS	

HIGH TEMP									
LOW TEMP									

SCHEDULE INFORMATION	INSPECTION
Date of projected completion: / /	
Is project on schedule: ☐ Yes ☐ No	
Days behind:	

JOBSITE NOTES

INJURIES	
Injuries on the job: ☐ Yes ☐ No	
If yes, was OSHA notified: ☐ Yes ☐ No	
Type of injury: ☐ First Aid ☐ Hospital	
(Details MUST be listed under comments)	

SAFETY	
Toolbox topic:	Signage posted: ☐ Yes ☐ No
Everyone wearing PPE: ☐ Yes ☐ No	Checklist Complete: ☐ Yes ☐ No

TODAY'S DATE	JOB/PROJECT INFORMATION
INITIALS	

HIGH TEMP		
LOW TEMP		

SCHEDULE INFORMATION	INSPECTION
Date of projected completion: / /	
Is project on schedule: ☐ Yes ☐ No	
Days behind:	

JOBSITE NOTES

INJURIES		SAFETY	
Injuries on the job: ☐ Yes ☐ No		Toolbox topic:	Signage posted: ☐ Yes ☐ No
If yes, was OSHA notified: ☐ Yes ☐ No			
Type of injury: ☐ First Aid ☐ Hospital (*Details MUST be listed under comments*)		Everyone wearing PPE: ☐ Yes ☐ No	Checklist Complete: ☐ Yes ☐ No

TODAY'S DATE	JOB/PROJECT INFORMATION

INITIALS	

HIGH TEMP									
LOW TEMP									

SCHEDULE INFORMATION	INSPECTION
Date of projected completion: / /	
Is project on schedule: ☐ Yes ☐ No	
Days behind:	

JOBSITE NOTES

INJURIES		SAFETY	
Injuries on the job: ☐ Yes ☐ No		Toolbox topic:	Signage posted: ☐ Yes ☐ No
If yes, was OSHA notified: ☐ Yes ☐ No			
Type of injury: ☐ First Aid ☐ Hospital (*Details MUST be listed under comments*)		Everyone wearing PPE: ☐ Yes ☐ No	Checklist Complete: ☐ Yes ☐ No

TODAY'S DATE	JOB/PROJECT INFORMATION
INITIALS	

HIGH TEMP LOW TEMP									

SCHEDULE INFORMATION	INSPECTION
Date of projected completion: / /	
Is project on schedule: ☐ Yes ☐ No	
Days behind:	

JOBSITE NOTES

INJURIES

Injuries on the job:	☐ Yes	☐ No
If yes, was OSHA notified:	☐ Yes	☐ No
Type of injury:	☐ First Aid	☐ Hospital

(*Details MUST be listed under comments*)

SAFETY

Toolbox topic:	Signage posted: ☐ Yes ☐ No
Everyone wearing PPE: ☐ Yes ☐ No	Checklist Complete: ☐ Yes ☐ No

TODAY'S DATE	JOB/PROJECT INFORMATION
INITIALS	

HIGH TEMP								
LOW TEMP								

SCHEDULE INFORMATION	INSPECTION
Date of projected completion: / /	
Is project on schedule: ☐ Yes ☐ No	
Days behind:	

JOBSITE NOTES

_____

_____

_____

_____

_____

_____

_____

_____

_____

_____

_____

_____

_____

_____

_____

_____

_____

INJURIES		SAFETY	
Injuries on the job: ☐ Yes ☐ No		Toolbox topic:	Signage posted: ☐ Yes ☐ No
If yes, was OSHA notified: ☐ Yes ☐ No		Everyone wearing PPE: ☐ Yes ☐ No	Checklist Complete: ☐ Yes ☐ No
Type of injury: ☐ First Aid ☐ Hospital _(Details MUST be listed under comments)_			

TODAY'S DATE	JOB/PROJECT INFORMATION
INITIALS	

HIGH TEMP		
LOW TEMP		

SCHEDULE INFORMATION	INSPECTION
Date of projected completion: / /	
Is project on schedule: ☐ Yes ☐ No	
Days behind:	

JOBSITE NOTES

INJURIES	
Injuries on the job: ☐ Yes ☐ No	
If yes, was OSHA notified: ☐ Yes ☐ No	
Type of injury: ☐ First Aid ☐ Hospital	
(*Details MUST be listed under comments*)	

SAFETY	
Toolbox topic:	Signage posted: ☐ Yes ☐ No
Everyone wearing PPE: ☐ Yes ☐ No	Checklist Complete: ☐ Yes ☐ No

TODAY'S DATE	JOB/PROJECT INFORMATION
INITIALS	

		HIGH TEMP								

HIGH TEMP	
LOW TEMP	

SCHEDULE INFORMATION	INSPECTION
Date of projected completion: / /	
Is project on schedule: ☐ Yes ☐ No	
Days behind:	

JOBSITE NOTES

INJURIES

Injuries on the job: ☐ Yes ☐ No

If yes, was OSHA notified: ☐ Yes ☐ No

Type of injury: ☐ First Aid ☐ Hospital
(*Details MUST be listed under comments*)

SAFETY

Toolbox topic:	Signage posted: ☐ Yes ☐ No
Everyone wearing PPE: ☐ Yes ☐ No	Checklist Complete: ☐ Yes ☐ No

TODAY'S DATE	JOB/PROJECT INFORMATION
INITIALS	

HIGH TEMP									
LOW TEMP									

SCHEDULE INFORMATION	INSPECTION
Date of projected completion: / /	
Is project on schedule: ☐ Yes ☐ No	
Days behind:	

JOBSITE NOTES

INJURIES		SAFETY	
Injuries on the job: ☐ Yes ☐ No		Toolbox topic:	Signage posted: ☐ Yes ☐ No
If yes, was OSHA notified: ☐ Yes ☐ No			
Type of injury: ☐ First Aid ☐ Hospital		Everyone wearing PPE: ☐ Yes ☐ No	Checklist Complete: ☐ Yes ☐ No
(*Details MUST be listed under comments*)			

TODAY'S DATE	JOB/PROJECT INFORMATION
INITIALS	

HIGH TEMP									
LOW TEMP									

SCHEDULE INFORMATION	INSPECTION
Date of projected completion: / /	
Is project on schedule: ☐ Yes ☐ No	
Days behind:	

JOBSITE NOTES

INJURIES	SAFETY	
Injuries on the job: ☐ Yes ☐ No	Toolbox topic:	Signage posted: ☐ Yes ☐ No
If yes, was OSHA notified: ☐ Yes ☐ No		
Type of injury: ☐ First Aid ☐ Hospital *(Details MUST be listed under comments)*	Everyone wearing PPE: ☐ Yes ☐ No	Checklist Complete: ☐ Yes ☐ No

MAIN LOG

TODAY'S DATE	JOB/PROJECT INFORMATION
INITIALS	

HIGH TEMP		
LOW TEMP		

SCHEDULE INFORMATION	INSPECTION
Date of projected completion: / /	
Is project on schedule: ☐ Yes ☐ No	
Days behind:	

JOBSITE NOTES

INJURIES

Injuries on the job: ☐ Yes ☐ No

If yes, was OSHA notified: ☐ Yes ☐ No

Type of injury: ☐ First Aid ☐ Hospital
(*Details MUST be listed under comments*)

SAFETY

Toolbox topic:	Signage posted: ☐ Yes ☐ No
Everyone wearing PPE: ☐ Yes ☐ No	Checklist Complete: ☐ Yes ☐ No

TODAY'S DATE	JOB/PROJECT INFORMATION
INITIALS	

HIGH TEMP									
LOW TEMP									

SCHEDULE INFORMATION	INSPECTION
Date of projected completion: / /	
Is project on schedule: ☐ Yes ☐ No	
Days behind:	

JOBSITE NOTES

INJURIES	SAFETY	
Injuries on the job: ☐ Yes ☐ No	Toolbox topic:	Signage posted: ☐ Yes ☐ No
If yes, was OSHA notified: ☐ Yes ☐ No		
Type of injury: ☐ First Aid ☐ Hospital	Everyone wearing PPE: ☐ Yes ☐ No	Checklist Complete: ☐ Yes ☐ No
(Details MUST be listed under comments)		

TODAY'S DATE	JOB/PROJECT INFORMATION
INITIALS	

HIGH TEMP									
LOW TEMP									

SCHEDULE INFORMATION	INSPECTION
Date of projected completion: / /	
Is project on schedule: ☐ Yes ☐ No	
Days behind:	

JOBSITE NOTES

INJURIES		SAFETY	
Injuries on the job: ☐ Yes ☐ No		Toolbox topic:	Signage posted: ☐ Yes ☐ No
If yes, was OSHA notified: ☐ Yes ☐ No			
Type of injury: ☐ First Aid ☐ Hospital (*Details MUST be listed under comments*)		Everyone wearing PPE: ☐ Yes ☐ No	Checklist Complete: ☐ Yes ☐ No

TODAY'S DATE	JOB/PROJECT INFORMATION
INITIALS	

HIGH TEMP								
LOW TEMP								

SCHEDULE INFORMATION	INSPECTION
Date of projected completion: / /	
Is project on schedule: ☐ Yes ☐ No	
Days behind:	

JOBSITE NOTES

INJURIES	
Injuries on the job: ☐ Yes ☐ No	
If yes, was OSHA notified: ☐ Yes ☐ No	
Type of injury: ☐ First Aid ☐ Hospital *(Details MUST be listed under comments)*	

SAFETY	
Toolbox topic:	Signage posted: ☐ Yes ☐ No
Everyone wearing PPE: ☐ Yes ☐ No	Checklist Complete: ☐ Yes ☐ No

TODAY'S DATE	JOB/PROJECT INFORMATION
INITIALS	

HIGH TEMP									
LOW TEMP									

SCHEDULE INFORMATION	INSPECTION
Date of projected completion: / /	
Is project on schedule: ☐ Yes ☐ No	
Days behind:	

JOBSITE NOTES

_____

_____

_____

_____

_____

_____

_____

_____

_____

_____

_____

_____

_____

_____

_____

_____

_____

INJURIES

Injuries on the job: ☐ Yes   ☐ No

If yes, was OSHA notified: ☐ Yes   ☐ No

Type of injury: ☐ First Aid   ☐ Hospital
(_Details MUST be listed under comments_)

SAFETY

Toolbox topic:	Signage posted: ☐ Yes ☐ No
Everyone wearing PPE: ☐ Yes ☐ No	Checklist Complete: ☐ Yes ☐ No

TODAY'S DATE	JOB/PROJECT INFORMATION
INITIALS	

HIGH TEMP **LOW TEMP**									

SCHEDULE INFORMATION	INSPECTION
Date of projected completion: / /	
Is project on schedule: ☐ Yes ☐ No	
Days behind:	

JOBSITE NOTES

INJURIES		SAFETY	
Injuries on the job: ☐ Yes ☐ No		Toolbox topic:	Signage posted: ☐ Yes ☐ No
If yes, was OSHA notified: ☐ Yes ☐ No			
Type of injury: ☐ First Aid ☐ Hospital *(Details MUST be listed under comments)*		Everyone wearing PPE: ☐ Yes ☐ No	Checklist Complete: ☐ Yes ☐ No

TODAY'S DATE	JOB/PROJECT INFORMATION

INITIALS	

HIGH TEMP		
LOW TEMP		

SCHEDULE INFORMATION	INSPECTION
Date of projected completion: / /	
Is project on schedule: ☐ Yes ☐ No	
Days behind:	

JOBSITE NOTES

INJURIES

Injuries on the job: ☐ Yes   ☐ No

If yes, was OSHA notified: ☐ Yes   ☐ No

Type of injury: ☐ First Aid   ☐ Hospital
(*Details MUST be listed under comments*)

SAFETY

Toolbox topic:	Signage posted: ☐ Yes ☐ No
Everyone wearing PPE: ☐ Yes ☐ No	Checklist Complete: ☐ Yes ☐ No

TODAY'S DATE	JOB/PROJECT INFORMATION
INITIALS	

HIGH TEMP LOW TEMP								

SCHEDULE INFORMATION	INSPECTION
Date of projected completion: / /	
Is project on schedule: ☐ Yes ☐ No	
Days behind:	

JOBSITE NOTES

INJURIES

Injuries on the job: ☐ Yes ☐ No		
If yes, was OSHA notified: ☐ Yes ☐ No		
Type of injury: ☐ First Aid ☐ Hospital		
(Details MUST be listed under comments)		

SAFETY

Toolbox topic:	Signage posted: ☐ Yes ☐ No
Everyone wearing PPE: ☐ Yes ☐ No	Checklist Complete: ☐ Yes ☐ No

TODAY'S DATE	JOB/PROJECT INFORMATION
INITIALS	

HIGH TEMP									
LOW TEMP									

SCHEDULE INFORMATION	INSPECTION
Date of projected completion: / /	
Is project on schedule: ☐ Yes ☐ No	
Days behind:	

JOBSITE NOTES

_____

INJURIES

Injuries on the job: ☐ Yes   ☐ No

If yes, was OSHA notified: ☐ Yes   ☐ No

Type of injury: ☐ First Aid   ☐ Hospital
(*Details MUST be listed under comments*)

SAFETY

Toolbox topic:	Signage posted: ☐ Yes ☐ No
Everyone wearing PPE: ☐ Yes ☐ No	Checklist Complete: ☐ Yes ☐ No

TODAY'S DATE	JOB/PROJECT INFORMATION
INITIALS	

HIGH TEMP									
LOW TEMP									

SCHEDULE INFORMATION	INSPECTION
Date of projected completion: / /	
Is project on schedule: ☐ Yes ☐ No	
Days behind:	

JOBSITE NOTES

INJURIES			SAFETY	
Injuries on the job: ☐ Yes ☐ No			Toolbox topic:	Signage posted: ☐ Yes ☐ No
If yes, was OSHA notified: ☐ Yes ☐ No			Everyone wearing PPE: ☐ Yes ☐ No	Checklist Complete: ☐ Yes ☐ No
Type of injury: ☐ First Aid ☐ Hospital *(Details MUST be listed under comments)*				

TODAY'S DATE	JOB/PROJECT INFORMATION
INITIALS	

HIGH TEMP LOW TEMP								

SCHEDULE INFORMATION	INSPECTION
Date of projected completion: / /	
Is project on schedule: ☐ Yes ☐ No	
Days behind:	

JOBSITE NOTES

INJURIES

Injuries on the job: ☐ Yes   ☐ No

If yes, was OSHA notified: ☐ Yes   ☐ No

Type of injury: ☐ First Aid   ☐ Hospital
(*Details MUST be listed under comments*)

SAFETY

Toolbox topic:	Signage posted: ☐ Yes ☐ No
Everyone wearing PPE: ☐ Yes ☐ No	Checklist Complete: ☐ Yes ☐ No

TODAY'S DATE	JOB/PROJECT INFORMATION
INITIALS	

HIGH TEMP									
LOW TEMP									

SCHEDULE INFORMATION	INSPECTION
Date of projected completion: / /	
Is project on schedule: ☐ Yes ☐ No	
Days behind:	

JOBSITE NOTES

INJURIES			SAFETY	
Injuries on the job: ☐ Yes ☐ No			Toolbox topic:	Signage posted: ☐ Yes ☐ No
If yes, was OSHA notified: ☐ Yes ☐ No				
Type of injury: ☐ First Aid ☐ Hospital			Everyone wearing PPE: ☐ Yes ☐ No	Checklist Complete: ☐ Yes ☐ No
(Details MUST be listed under comments)				

TODAY'S DATE	JOB/PROJECT INFORMATION
INITIALS	

HIGH TEMP									
LOW TEMP									

SCHEDULE INFORMATION	INSPECTION
Date of projected completion: / /	
Is project on schedule: ☐ Yes ☐ No	
Days behind:	

JOBSITE NOTES

INJURIES	
Injuries on the job: ☐ Yes ☐ No	
If yes, was OSHA notified: ☐ Yes ☐ No	
Type of injury: ☐ First Aid ☐ Hospital	
(*Details MUST be listed under comments*)	

SAFETY	
Toolbox topic:	Signage posted: ☐ Yes ☐ No
Everyone wearing PPE: ☐ Yes ☐ No	Checklist Complete: ☐ Yes ☐ No

TODAY'S DATE	JOB/PROJECT INFORMATION
INITIALS	

HIGH TEMP									
LOW TEMP									

SCHEDULE INFORMATION	INSPECTION
Date of projected completion: / /	
Is project on schedule: ☐ Yes ☐ No	
Days behind:	

JOBSITE NOTES

INJURIES	SAFETY	
Injuries on the job: ☐ Yes ☐ No	Toolbox topic:	Signage posted: ☐ Yes ☐ No
If yes, was OSHA notified: ☐ Yes ☐ No		
Type of injury: ☐ First Aid ☐ Hospital	Everyone wearing PPE: ☐ Yes ☐ No	Checklist Complete: ☐ Yes ☐ No
(Details MUST be listed under comments)		

TODAY'S DATE	JOB/PROJECT INFORMATION
INITIALS	

HIGH TEMP LOW TEMP									

SCHEDULE INFORMATION	INSPECTION
Date of projected completion: / /	
Is project on schedule: ☐ Yes ☐ No	
Days behind:	

JOBSITE NOTES

INJURIES

Injuries on the job: ☐ Yes ☐ No	
If yes, was OSHA notified: ☐ Yes ☐ No	
Type of injury: ☐ First Aid ☐ Hospital	
(*Details MUST be listed under comments*)	

SAFETY

Toolbox topic:	Signage posted: ☐ Yes ☐ No
Everyone wearing PPE: ☐ Yes ☐ No	Checklist Complete: ☐ Yes ☐ No

TODAY'S DATE	JOB/PROJECT INFORMATION
INITIALS	

HIGH TEMP										
LOW TEMP										

SCHEDULE INFORMATION	INSPECTION
Date of projected completion: / /	
Is project on schedule: ☐ Yes ☐ No	
Days behind:	

JOBSITE NOTES

INJURIES	SAFETY	
Injuries on the job: ☐ Yes ☐ No	Toolbox topic:	Signage posted: ☐ Yes ☐ No
If yes, was OSHA notified: ☐ Yes ☐ No		
Type of injury: ☐ First Aid ☐ Hospital	Everyone wearing PPE: ☐ Yes ☐ No	Checklist Complete: ☐ Yes ☐ No
(*Details MUST be listed under comments*)		

TODAY'S DATE	JOB/PROJECT INFORMATION
INITIALS	

HIGH TEMP	
LOW TEMP	

SCHEDULE INFORMATION	INSPECTION
Date of projected completion: / /	
Is project on schedule: ☐ Yes ☐ No	
Days behind:	

JOBSITE NOTES

INJURIES

Injuries on the job:	☐ Yes	☐ No
If yes, was OSHA notified:	☐ Yes	☐ No
Type of injury:	☐ First Aid	☐ Hospital

(*Details MUST be listed under comments*)

SAFETY

Toolbox topic:	Signage posted: ☐ Yes ☐ No
Everyone wearing PPE: ☐ Yes ☐ No	Checklist Complete: ☐ Yes ☐ No

TODAY'S DATE	JOB/PROJECT INFORMATION
INITIALS	

		HIGH TEMP								
HIGH TEMP										
LOW TEMP										

SCHEDULE INFORMATION	INSPECTION
Date of projected completion: / /	
Is project on schedule: ☐ Yes ☐ No	
Days behind:	

JOBSITE NOTES

(blank lined area)

INJURIES	SAFETY	
Injuries on the job: ☐ Yes ☐ No	Toolbox topic:	Signage posted: ☐ Yes ☐ No
If yes, was OSHA notified: ☐ Yes ☐ No		
Type of injury: ☐ First Aid ☐ Hospital	Everyone wearing PPE: ☐ Yes ☐ No	Checklist Complete: ☐ Yes ☐ No
(_Details MUST be listed under comments_)		

TODAY'S DATE	JOB/PROJECT INFORMATION
INITIALS	

HIGH TEMP LOW TEMP								

SCHEDULE INFORMATION	INSPECTION
Date of projected completion: / /	
Is project on schedule: ☐ Yes ☐ No	
Days behind:	

JOBSITE NOTES

INJURIES
Injuries on the job: ☐ Yes   ☐ No
If yes, was OSHA notified: ☐ Yes   ☐ No
Type of injury: ☐ First Aid   ☐ Hospital
(Details MUST be listed under comments)

SAFETY	
Toolbox topic:	Signage posted: ☐ Yes ☐ No
Everyone wearing PPE: ☐ Yes ☐ No	Checklist Complete: ☐ Yes ☐ No

TODAY'S DATE	JOB/PROJECT INFORMATION
INITIALS	

HIGH TEMP									
LOW TEMP									

SCHEDULE INFORMATION	INSPECTION
Date of projected completion: / /	
Is project on schedule: ☐ Yes ☐ No	
Days behind:	

JOBSITE NOTES

INJURIES		SAFETY	
Injuries on the job: ☐ Yes ☐ No		Toolbox topic:	Signage posted: ☐ Yes ☐ No
If yes, was OSHA notified: ☐ Yes ☐ No			
Type of injury: ☐ First Aid ☐ Hospital		Everyone wearing PPE: ☐ Yes ☐ No	Checklist Complete: ☐ Yes ☐ No
(*Details MUST be listed under comments*)			

TODAY'S DATE	JOB/PROJECT INFORMATION
INITIALS	

HIGH TEMP									
LOW TEMP									

SCHEDULE INFORMATION	INSPECTION
Date of projected completion: / /	
Is project on schedule: ☐ Yes ☐ No	
Days behind:	

JOBSITE NOTES

_____
_____
_____
_____
_____
_____
_____
_____
_____
_____
_____
_____
_____
_____
_____
_____
_____

INJURIES	
Injuries on the job: ☐ Yes ☐ No	
If yes, was OSHA notified: ☐ Yes ☐ No	
Type of injury: ☐ First Aid ☐ Hospital	
(Details MUST be listed under comments)	

SAFETY	
Toolbox topic:	Signage posted: ☐ Yes ☐ No
Everyone wearing PPE: ☐ Yes ☐ No	Checklist Complete: ☐ Yes ☐ No

TODAY'S DATE	JOB/PROJECT INFORMATION
INITIALS	

HIGH TEMP									
LOW TEMP									

SCHEDULE INFORMATION	INSPECTION
Date of projected completion: / /	
Is project on schedule: ☐ Yes ☐ No	
Days behind:	

JOBSITE NOTES

INJURIES		SAFETY	
Injuries on the job: ☐ Yes ☐ No		Toolbox topic:	Signage posted: ☐ Yes ☐ No
If yes, was OSHA notified: ☐ Yes ☐ No			
Type of injury: ☐ First Aid ☐ Hospital *(Details MUST be listed under comments)*		Everyone wearing PPE: ☐ Yes ☐ No	Checklist Complete: ☐ Yes ☐ No

TODAY'S DATE	JOB/PROJECT INFORMATION
INITIALS	

HIGH TEMP									
LOW TEMP									

SCHEDULE INFORMATION	INSPECTION
Date of projected completion: / /	
Is project on schedule: ☐ Yes ☐ No	
Days behind:	

JOBSITE NOTES

INJURIES		SAFETY	
Injuries on the job: ☐ Yes ☐ No		Toolbox topic:	Signage posted: ☐ Yes ☐ No
If yes, was OSHA notified: ☐ Yes ☐ No			
Type of injury: ☐ First Aid ☐ Hospital (*Details MUST be listed under comments*)		Everyone wearing PPE: ☐ Yes ☐ No	Checklist Complete: ☐ Yes ☐ No

TODAY'S DATE	JOB/PROJECT INFORMATION
INITIALS	

HIGH TEMP									
LOW TEMP									

SCHEDULE INFORMATION	INSPECTION
Date of projected completion: / /	
Is project on schedule: ☐ Yes ☐ No	
Days behind:	

JOBSITE NOTES

_____

_____

_____

_____

_____

_____

_____

_____

_____

_____

_____

_____

_____

_____

_____

_____

INJURIES	SAFETY	
Injuries on the job: ☐ Yes ☐ No	Toolbox topic:	Signage posted: ☐ Yes ☐ No
If yes, was OSHA notified: ☐ Yes ☐ No		
Type of injury: ☐ First Aid ☐ Hospital	Everyone wearing PPE: ☐ Yes ☐ No	Checklist Complete: ☐ Yes ☐ No
(*Details MUST be listed under comments*)		

TODAY'S DATE	JOB/PROJECT INFORMATION
INITIALS	

HIGH TEMP									
LOW TEMP									

SCHEDULE INFORMATION	INSPECTION
Date of projected completion: / /	
Is project on schedule: ☐ Yes ☐ No	
Days behind:	

JOBSITE NOTES

INJURIES		SAFETY	
Injuries on the job: ☐ Yes ☐ No		Toolbox topic:	Signage posted: ☐ Yes ☐ No
If yes, was OSHA notified: ☐ Yes ☐ No			
Type of injury: ☐ First Aid ☐ Hospital (*Details MUST be listed under comments*)		Everyone wearing PPE: ☐ Yes ☐ No	Checklist Complete: ☐ Yes ☐ No

TODAY'S DATE	JOB/PROJECT INFORMATION
INITIALS	

HIGH TEMP									
LOW TEMP									

SCHEDULE INFORMATION	INSPECTION
Date of projected completion: / /	
Is project on schedule: ☐ Yes ☐ No	
Days behind:	

JOBSITE NOTES

INJURIES		
Injuries on the job: ☐ Yes ☐ No		
If yes, was OSHA notified: ☐ Yes ☐ No		
Type of injury: ☐ First Aid ☐ Hospital *(Details MUST be listed under comments)*		

SAFETY	
Toolbox topic:	Signage posted: ☐ Yes ☐ No
Everyone wearing PPE: ☐ Yes ☐ No	Checklist Complete: ☐ Yes ☐ No

TODAY'S DATE	JOB/PROJECT INFORMATION
INITIALS	

HIGH TEMP									
LOW TEMP									

SCHEDULE INFORMATION	INSPECTION
Date of projected completion: / /	
Is project on schedule: ☐ Yes ☐ No	
Days behind:	

JOBSITE NOTES

_____

_____

_____

_____

_____

_____

_____

_____

_____

_____

_____

_____

_____

_____

_____

INJURIES	SAFETY	
Injuries on the job: ☐ Yes ☐ No	Toolbox topic:	Signage posted: ☐ Yes ☐ No
If yes, was OSHA notified: ☐ Yes ☐ No		
Type of injury: ☐ First Aid ☐ Hospital (_Details MUST be listed under comments_)	Everyone wearing PPE: ☐ Yes ☐ No	Checklist Complete: ☐ Yes ☐ No

TODAY'S DATE	JOB/PROJECT INFORMATION

INITIALS	

HIGH TEMP	
LOW TEMP	

SCHEDULE INFORMATION	INSPECTION
Date of projected completion: / /	
Is project on schedule: ☐ Yes ☐ No	
Days behind:	

JOBSITE NOTES

INJURIES

Injuries on the job: ☐ Yes ☐ No

If yes, was OSHA notified: ☐ Yes ☐ No

Type of injury: ☐ First Aid ☐ Hospital
(*Details MUST be listed under comments*)

SAFETY

Toolbox topic:	Signage posted: ☐ Yes ☐ No
Everyone wearing PPE: ☐ Yes ☐ No	Checklist Complete: ☐ Yes ☐ No

TODAY'S DATE	JOB/PROJECT INFORMATION
INITIALS	

HIGH TEMP		
LOW TEMP		

SCHEDULE INFORMATION	INSPECTION
Date of projected completion: / /	
Is project on schedule: ☐ Yes ☐ No	
Days behind:	

JOBSITE NOTES

INJURIES		SAFETY	
Injuries on the job: ☐ Yes ☐ No		Toolbox topic:	Signage posted: ☐ Yes ☐ No
If yes, was OSHA notified: ☐ Yes ☐ No			
Type of injury: ☐ First Aid ☐ Hospital (*Details MUST be listed under comments*)		Everyone wearing PPE: ☐ Yes ☐ No	Checklist Complete: ☐ Yes ☐ No

TODAY'S DATE	JOB/PROJECT INFORMATION
INITIALS	

HIGH TEMP										
LOW TEMP										

SCHEDULE INFORMATION	INSPECTION
Date of projected completion: / /	
Is project on schedule: ☐ Yes ☐ No	
Days behind:	

JOBSITE NOTES

INJURIES		SAFETY	
Injuries on the job: ☐ Yes ☐ No		Toolbox topic:	Signage posted: ☐ Yes ☐ No
If yes, was OSHA notified: ☐ Yes ☐ No			
Type of injury: ☐ First Aid ☐ Hospital		Everyone wearing PPE: ☐ Yes ☐ No	Checklist Complete: ☐ Yes ☐ No
(*Details MUST be listed under comments*)			

TODAY'S DATE	JOB/PROJECT INFORMATION
INITIALS	

HIGH TEMP									
LOW TEMP									

SCHEDULE INFORMATION	INSPECTION
Date of projected completion: / /	
Is project on schedule: ☐ Yes ☐ No	
Days behind:	

JOBSITE NOTES

INJURIES			SAFETY	
Injuries on the job: ☐ Yes ☐ No			Toolbox topic:	Signage posted: ☐ Yes ☐ No
If yes, was OSHA notified: ☐ Yes ☐ No				
Type of injury: ☐ First Aid ☐ Hospital			Everyone wearing PPE: ☐ Yes ☐ No	Checklist Complete: ☐ Yes ☐ No
(*Details MUST be listed under comments*)				

TODAY'S DATE	JOB/PROJECT INFORMATION
INITIALS	

		HIGH TEMP							
HIGH TEMP									
LOW TEMP									

SCHEDULE INFORMATION	INSPECTION
Date of projected completion: / /	
Is project on schedule: ☐ Yes ☐ No	
Days behind:	

JOBSITE NOTES

INJURIES

Injuries on the job: ☐ Yes   ☐ No

If yes, was OSHA notified: ☐ Yes   ☐ No

Type of injury: ☐ First Aid   ☐ Hospital

(*Details MUST be listed under comments*)

SAFETY

Toolbox topic:	Signage posted: ☐ Yes ☐ No
Everyone wearing PPE: ☐ Yes ☐ No	Checklist Complete: ☐ Yes ☐ No

TODAY'S DATE	JOB/PROJECT INFORMATION
INITIALS	

HIGH TEMP										
LOW TEMP										

SCHEDULE INFORMATION	INSPECTION
Date of projected completion: / /	
Is project on schedule: ☐ Yes ☐ No	
Days behind:	

JOBSITE NOTES

INJURIES

Injuries on the job: ☐ Yes ☐ No

If yes, was OSHA notified: ☐ Yes ☐ No

Type of injury: ☐ First Aid ☐ Hospital
(*Details MUST be listed under comments*)

SAFETY

Toolbox topic:	Signage posted: ☐ Yes ☐ No
Everyone wearing PPE: ☐ Yes ☐ No	Checklist Complete: ☐ Yes ☐ No

TODAY'S DATE	JOB/PROJECT INFORMATION
INITIALS	

		HIGH TEMP		

HIGH TEMP	
LOW TEMP	

SCHEDULE INFORMATION	INSPECTION
Date of projected completion: / /	
Is project on schedule: ☐ Yes ☐ No	
Days behind:	

JOBSITE NOTES

INJURIES	SAFETY	
Injuries on the job: ☐ Yes ☐ No	Toolbox topic:	Signage posted: ☐ Yes ☐ No
If yes, was OSHA notified: ☐ Yes ☐ No		
Type of injury: ☐ First Aid ☐ Hospital	Everyone wearing PPE: ☐ Yes ☐ No	Checklist Complete: ☐ Yes ☐ No
(*Details MUST be listed under comments*)		

MAIN LOG

TODAY'S DATE	JOB/PROJECT INFORMATION
INITIALS	

HIGH TEMP									
LOW TEMP									

SCHEDULE INFORMATION	INSPECTION
Date of projected completion: / /	
Is project on schedule: ☐ Yes ☐ No	
Days behind:	

JOBSITE NOTES

_____
_____
_____
_____
_____
_____
_____
_____
_____
_____
_____
_____
_____
_____
_____
_____

INJURIES

Injuries on the job: ☐ Yes   ☐ No

If yes, was OSHA notified: ☐ Yes   ☐ No

Type of injury: ☐ First Aid   ☐ Hospital
(*Details MUST be listed under comments*)

SAFETY

Toolbox topic:	Signage posted: ☐ Yes ☐ No
Everyone wearing PPE: ☐ Yes ☐ No	Checklist Complete: ☐ Yes ☐ No

TODAY'S DATE	JOB/PROJECT INFORMATION
INITIALS	

HIGH TEMP									
LOW TEMP									

SCHEDULE INFORMATION	INSPECTION
Date of projected completion: / /	
Is project on schedule: ☐Yes ☐No	
Days behind:	

JOBSITE NOTES

_____

_____

_____

_____

_____

_____

_____

_____

_____

_____

_____

_____

_____

_____

_____

_____

INJURIES	SAFETY	
Injuries on the job: ☐Yes ☐No	Toolbox topic:	Signage posted: ☐Yes ☐No
If yes, was OSHA notified: ☐Yes ☐No		
Type of injury: ☐First Aid ☐Hospital _(Details MUST be listed under comments)_	Everyone wearing PPE: ☐Yes ☐No	Checklist Complete: ☐Yes ☐No

TODAY'S DATE	JOB/PROJECT INFORMATION
INITIALS	

HIGH TEMP									
LOW TEMP									

SCHEDULE INFORMATION

SCHEDULE INFORMATION	INSPECTION
Date of projected completion: / /	
Is project on schedule: ☐ Yes ☐ No	
Days behind:	

JOBSITE NOTES

INJURIES

INJURIES
Injuries on the job: ☐ Yes   ☐ No
If yes, was OSHA notified: ☐ Yes   ☐ No
Type of injury: ☐ First Aid   ☐ Hospital
(Details MUST be listed under comments)

SAFETY

Toolbox topic:	Signage posted: ☐ Yes ☐ No
Everyone wearing PPE: ☐ Yes ☐ No	Checklist Complete: ☐ Yes ☐ No

TODAY'S DATE	JOB/PROJECT INFORMATION
INITIALS	

HIGH TEMP									
LOW TEMP									

SCHEDULE INFORMATION	INSPECTION
Date of projected completion: / /	
Is project on schedule: ☐ Yes ☐ No	
Days behind:	

JOBSITE NOTES

INJURIES		
Injuries on the job: ☐ Yes ☐ No		
If yes, was OSHA notified: ☐ Yes ☐ No		
Type of injury: ☐ First Aid ☐ Hospital *(Details MUST be listed under comments)*		

SAFETY	
Toolbox topic:	Signage posted: ☐ Yes ☐ No
Everyone wearing PPE: ☐ Yes ☐ No	Checklist Complete: ☐ Yes ☐ No

TODAY'S DATE	JOB/PROJECT INFORMATION
INITIALS	

HIGH TEMP		
LOW TEMP		

SCHEDULE INFORMATION	INSPECTION
Date of projected completion: / /	
Is project on schedule: ☐ Yes ☐ No	
Days behind:	

JOBSITE NOTES

INJURIES

Injuries on the job: ☐ Yes ☐ No	
If yes, was OSHA notified: ☐ Yes ☐ No	
Type of injury: ☐ First Aid ☐ Hospital	
(*Details MUST be listed under comments*)	

SAFETY

Toolbox topic:	Signage posted: ☐ Yes ☐ No
Everyone wearing PPE: ☐ Yes ☐ No	Checklist Complete: ☐ Yes ☐ No

TODAY'S DATE	JOB/PROJECT INFORMATION
INITIALS	

HIGH TEMP									
LOW TEMP									

SCHEDULE INFORMATION	INSPECTION
Date of projected completion: / /	
Is project on schedule: ☐ Yes ☐ No	
Days behind:	

JOBSITE NOTES

INJURIES		SAFETY	
Injuries on the job: ☐ Yes ☐ No		Toolbox topic:	Signage posted: ☐ Yes ☐ No
If yes, was OSHA notified: ☐ Yes ☐ No			
Type of injury: ☐ First Aid ☐ Hospital		Everyone wearing PPE: ☐ Yes ☐ No	Checklist Complete: ☐ Yes ☐ No
(*Details MUST be listed under comments*)			

TODAY'S DATE	JOB/PROJECT INFORMATION
INITIALS	

HIGH TEMP									
LOW TEMP									

SCHEDULE INFORMATION	INSPECTION
Date of projected completion: / /	
Is project on schedule: ☐ Yes ☐ No	
Days behind:	

JOBSITE NOTES

INJURIES

Injuries on the job: ☐ Yes ☐ No	
If yes, was OSHA notified: ☐ Yes ☐ No	
Type of injury: ☐ First Aid ☐ Hospital	
(*Details MUST be listed under comments*)	

SAFETY

Toolbox topic:	Signage posted: ☐ Yes ☐ No
Everyone wearing PPE: ☐ Yes ☐ No	Checklist Complete: ☐ Yes ☐ No

TODAY'S DATE	JOB/PROJECT INFORMATION
INITIALS	

HIGH TEMP LOW TEMP									

SCHEDULE INFORMATION	INSPECTION
Date of projected completion: / /	
Is project on schedule: ☐ Yes ☐ No	
Days behind:	

JOBSITE NOTES

INJURIES	SAFETY	
Injuries on the job: ☐ Yes ☐ No	Toolbox topic:	Signage posted: ☐ Yes ☐ No
If yes, was OSHA notified: ☐ Yes ☐ No		
Type of injury: ☐ First Aid ☐ Hospital (*Details MUST be listed under comments*)	Everyone wearing PPE: ☐ Yes ☐ No	Checklist Complete: ☐ Yes ☐ No

TODAY'S DATE	JOB/PROJECT INFORMATION
INITIALS	

HIGH TEMP		
LOW TEMP		

SCHEDULE INFORMATION	INSPECTION
Date of projected completion: / /	
Is project on schedule: ☐ Yes ☐ No	
Days behind:	

JOBSITE NOTES

_____
_____
_____
_____
_____
_____
_____
_____
_____
_____
_____
_____
_____
_____
_____
_____
_____

INJURIES

Injuries on the job: ☐ Yes ☐ No

If yes, was OSHA notified: ☐ Yes ☐ No

Type of injury: ☐ First Aid ☐ Hospital
(*Details MUST be listed under comments*)

SAFETY

Toolbox topic:	Signage posted: ☐ Yes ☐ No
Everyone wearing PPE: ☐ Yes ☐ No	Checklist Complete: ☐ Yes ☐ No

TODAY'S DATE	JOB/PROJECT INFORMATION
INITIALS	

HIGH TEMP									
LOW TEMP									

SCHEDULE INFORMATION	INSPECTION
Date of projected completion: / /	
Is project on schedule: ☐ Yes ☐ No	
Days behind:	

JOBSITE NOTES

_____
_____
_____
_____
_____
_____
_____
_____
_____
_____
_____
_____
_____
_____
_____
_____
_____
_____

INJURIES	SAFETY	
Injuries on the job: ☐ Yes ☐ No	Toolbox topic:	Signage posted: ☐ Yes ☐ No
If yes, was OSHA notified: ☐ Yes ☐ No		
Type of injury: ☐ First Aid ☐ Hospital _(Details MUST be listed under comments)_	Everyone wearing PPE: ☐ Yes ☐ No	Checklist Complete: ☐ Yes ☐ No

TODAY'S DATE	JOB/PROJECT INFORMATION
INITIALS	

HIGH TEMP									
LOW TEMP									

SCHEDULE INFORMATION	INSPECTION
Date of projected completion: / /	
Is project on schedule: ☐ Yes ☐ No	
Days behind:	

JOBSITE NOTES

INJURIES	SAFETY	
Injuries on the job: ☐ Yes ☐ No	Toolbox topic:	Signage posted: ☐ Yes ☐ No
If yes, was OSHA notified: ☐ Yes ☐ No		
Type of injury: ☐ First Aid ☐ Hospital (*Details MUST be listed under comments*)	Everyone wearing PPE: ☐ Yes ☐ No	Checklist Complete: ☐ Yes ☐ No

TODAY'S DATE	JOB/PROJECT INFORMATION
INITIALS	

		HIGH TEMP		LOW TEMP			

HIGH TEMP	
LOW TEMP	

SCHEDULE INFORMATION	INSPECTION
Date of projected completion: / /	
Is project on schedule: ☐ Yes ☐ No	
Days behind:	

JOBSITE NOTES

INJURIES	SAFETY	
Injuries on the job: ☐ Yes ☐ No	Toolbox topic:	Signage posted: ☐ Yes ☐ No
If yes, was OSHA notified: ☐ Yes ☐ No		
Type of injury: ☐ First Aid ☐ Hospital	Everyone wearing PPE: ☐ Yes ☐ No	Checklist Complete: ☐ Yes ☐ No
(Details MUST be listed under comments)		

TODAY'S DATE	JOB/PROJECT INFORMATION
INITIALS	

HIGH TEMP		
LOW TEMP		

SCHEDULE INFORMATION	INSPECTION
Date of projected completion: / /	
Is project on schedule: ☐ Yes ☐ No	
Days behind:	

JOBSITE NOTES

INJURIES		SAFETY	
Injuries on the job: ☐ Yes ☐ No		Toolbox topic:	Signage posted: ☐ Yes ☐ No
If yes, was OSHA notified: ☐ Yes ☐ No			
Type of injury: ☐ First Aid ☐ Hospital		Everyone wearing PPE: ☐ Yes ☐ No	Checklist Complete: ☐ Yes ☐ No
(*Details MUST be listed under comments*)			

TODAY'S DATE	JOB/PROJECT INFORMATION
INITIALS	

HIGH TEMP									
LOW TEMP									

SCHEDULE INFORMATION	INSPECTION
Date of projected completion: / /	
Is project on schedule: ☐ Yes ☐ No	
Days behind:	

JOBSITE NOTES

_____

_____

_____

_____

_____

_____

_____

_____

_____

_____

_____

_____

_____

_____

_____

_____

_____

_____

INJURIES	SAFETY	
Injuries on the job: ☐ Yes ☐ No	Toolbox topic:	Signage posted: ☐ Yes ☐ No
If yes, was OSHA notified: ☐ Yes ☐ No		
Type of injury: ☐ First Aid ☐ Hospital _(Details MUST be listed under comments)_	Everyone wearing PPE: ☐ Yes ☐ No	Checklist Complete: ☐ Yes ☐ No

TODAY'S DATE	JOB/PROJECT INFORMATION
INITIALS	

HIGH TEMP	
LOW TEMP	

SCHEDULE INFORMATION	INSPECTION
Date of projected completion: / /	
Is project on schedule: ☐ Yes ☐ No	
Days behind:	

JOBSITE NOTES

INJURIES
Injuries on the job: ☐ Yes ☐ No
If yes, was OSHA notified: ☐ Yes ☐ No
Type of injury: ☐ First Aid ☐ Hospital (*Details MUST be listed under comments*)

SAFETY	
Toolbox topic:	Signage posted: ☐ Yes ☐ No
Everyone wearing PPE: ☐ Yes ☐ No	Checklist Complete: ☐ Yes ☐ No

TODAY'S DATE	JOB/PROJECT INFORMATION
INITIALS	

HIGH TEMP									
LOW TEMP									

SCHEDULE INFORMATION	INSPECTION
Date of projected completion: / /	
Is project on schedule: ☐ Yes ☐ No	
Days behind:	

JOBSITE NOTES

_____
_____
_____
_____
_____
_____
_____
_____
_____
_____
_____
_____
_____
_____
_____
_____
_____
_____

INJURIES	
Injuries on the job: ☐ Yes ☐ No	
If yes, was OSHA notified: ☐ Yes ☐ No	
Type of injury: ☐ First Aid ☐ Hospital	
(*Details MUST be listed under comments*)	

SAFETY	
Toolbox topic:	Signage posted: ☐ Yes ☐ No
Everyone wearing PPE: ☐ Yes ☐ No	Checklist Complete: ☐ Yes ☐ No

TODAY'S DATE	JOB/PROJECT INFORMATION
INITIALS	

HIGH TEMP	
LOW TEMP	

SCHEDULE INFORMATION	INSPECTION
Date of projected completion: / /	
Is project on schedule: ☐ Yes ☐ No	
Days behind:	

JOBSITE NOTES

INJURIES
Injuries on the job: ☐ Yes ☐ No
If yes, was OSHA notified: ☐ Yes ☐ No
Type of injury: ☐ First Aid ☐ Hospital
(Details MUST be listed under comments)

SAFETY	
Toolbox topic:	Signage posted: ☐ Yes ☐ No
Everyone wearing PPE: ☐ Yes ☐ No	Checklist Complete: ☐ Yes ☐ No

TODAY'S DATE	JOB/PROJECT INFORMATION
INITIALS	

HIGH TEMP									
LOW TEMP									

SCHEDULE INFORMATION	INSPECTION
Date of projected completion: / /	
Is project on schedule: ☐ Yes ☐ No	
Days behind:	

JOBSITE NOTES

INJURIES		SAFETY	
Injuries on the job: ☐ Yes ☐ No		Toolbox topic:	Signage posted: ☐ Yes ☐ No
If yes, was OSHA notified: ☐ Yes ☐ No			
Type of injury: ☐ First Aid ☐ Hospital *(Details MUST be listed under comments)*		Everyone wearing PPE: ☐ Yes ☐ No	Checklist Complete: ☐ Yes ☐ No

TODAY'S DATE	JOB/PROJECT INFORMATION
INITIALS	

HIGH TEMP LOW TEMP									

SCHEDULE INFORMATION	INSPECTION
Date of projected completion: / /	
Is project on schedule: ☐ Yes ☐ No	
Days behind:	

JOBSITE NOTES

INJURIES	SAFETY	
Injuries on the job: ☐ Yes ☐ No	Toolbox topic:	Signage posted: ☐ Yes ☐ No
If yes, was OSHA notified: ☐ Yes ☐ No		
Type of injury: ☐ First Aid ☐ Hospital	Everyone wearing PPE: ☐ Yes ☐ No	Checklist Complete: ☐ Yes ☐ No
(Details MUST be listed under comments)		

TODAY'S DATE	JOB/PROJECT INFORMATION
INITIALS	

HIGH TEMP									
LOW TEMP									

SCHEDULE INFORMATION	INSPECTION
Date of projected completion: / /	
Is project on schedule: ☐ Yes ☐ No	
Days behind:	

JOBSITE NOTES

INJURIES		SAFETY	
Injuries on the job: ☐ Yes ☐ No		Toolbox topic:	Signage posted: ☐ Yes ☐ No
If yes, was OSHA notified: ☐ Yes ☐ No			
Type of injury: ☐ First Aid ☐ Hospital		Everyone wearing PPE: ☐ Yes ☐ No	Checklist Complete: ☐ Yes ☐ No
(*Details MUST be listed under comments*)			

TODAY'S DATE	JOB/PROJECT INFORMATION
INITIALS	

HIGH TEMP								
LOW TEMP								

SCHEDULE INFORMATION	INSPECTION
Date of projected completion: / /	
Is project on schedule: ☐ Yes ☐ No	
Days behind:	

JOBSITE NOTES

INJURIES		SAFETY	
Injuries on the job: ☐ Yes ☐ No		Toolbox topic:	Signage posted: ☐ Yes ☐ No
If yes, was OSHA notified: ☐ Yes ☐ No			
Type of injury: ☐ First Aid ☐ Hospital		Everyone wearing PPE: ☐ Yes ☐ No	Checklist Complete: ☐ Yes ☐ No
(*Details MUST be listed under comments*)			

TODAY'S DATE	JOB/PROJECT INFORMATION
INITIALS	

HIGH TEMP LOW TEMP									

SCHEDULE INFORMATION	INSPECTION
Date of projected completion: / /	
Is project on schedule: ☐ Yes ☐ No	
Days behind:	

JOBSITE NOTES

INJURIES	SAFETY	
Injuries on the job: ☐ Yes ☐ No	Toolbox topic:	Signage posted: ☐ Yes ☐ No
If yes, was OSHA notified: ☐ Yes ☐ No		
Type of injury: ☐ First Aid ☐ Hospital	Everyone wearing PPE: ☐ Yes ☐ No	Checklist Complete: ☐ Yes ☐ No
(*Details MUST be listed under comments*)		

TODAY'S DATE	JOB/PROJECT INFORMATION
INITIALS	

HIGH TEMP	
LOW TEMP	

SCHEDULE INFORMATION	INSPECTION
Date of projected completion: / /	
Is project on schedule: ☐ Yes ☐ No	
Days behind:	

JOBSITE NOTES

_____

_____

_____

_____

_____

_____

_____

_____

_____

_____

_____

_____

_____

_____

_____

_____

_____

INJURIES

Injuries on the job: ☐ Yes   ☐ No

If yes, was OSHA notified: ☐ Yes   ☐ No

Type of injury: ☐ First Aid   ☐ Hospital
(*Details MUST be listed under comments*)

SAFETY

Toolbox topic:	Signage posted: ☐ Yes ☐ No
Everyone wearing PPE: ☐ Yes ☐ No	Checklist Complete: ☐ Yes ☐ No

TODAY'S DATE	JOB/PROJECT INFORMATION
INITIALS	

HIGH TEMP LOW TEMP									

SCHEDULE INFORMATION	INSPECTION
Date of projected completion: / /	
Is project on schedule: ☐ Yes ☐ No	
Days behind:	

JOBSITE NOTES

INJURIES			SAFETY	
Injuries on the job: ☐ Yes ☐ No			Toolbox topic:	Signage posted: ☐ Yes ☐ No
If yes, was OSHA notified: ☐ Yes ☐ No				
Type of injury: ☐ First Aid ☐ Hospital (*Details MUST be listed under comments*)			Everyone wearing PPE: ☐ Yes ☐ No	Checklist Complete: ☐ Yes ☐ No

TODAY'S DATE	JOB/PROJECT INFORMATION

INITIALS

| HIGH TEMP | | | | | | | | | |
| LOW TEMP | | | | | | | | | |

SCHEDULE INFORMATION

Date of projected completion: /   /

Is project on schedule: ☐ Yes ☐ No

Days behind:

INSPECTION

JOBSITE NOTES

INJURIES

Injuries on the job: ☐ Yes   ☐ No

If yes, was OSHA notified: ☐ Yes   ☐ No

Type of injury: ☐ First Aid   ☐ Hospital
(*Details MUST be listed under comments*)

SAFETY

Toolbox topic:	Signage posted: ☐ Yes ☐ No
Everyone wearing PPE: ☐ Yes ☐ No	Checklist Complete: ☐ Yes ☐ No

TODAY'S DATE	JOB/PROJECT INFORMATION
INITIALS	

HIGH TEMP									
LOW TEMP									

SCHEDULE INFORMATION	INSPECTION
Date of projected completion: / /	
Is project on schedule: ☐ Yes ☐ No	
Days behind:	

JOBSITE NOTES

INJURIES	
Injuries on the job: ☐ Yes ☐ No	
If yes, was OSHA notified: ☐ Yes ☐ No	
Type of injury: ☐ First Aid ☐ Hospital	
(*Details MUST be listed under comments*)	

SAFETY	
Toolbox topic:	Signage posted: ☐ Yes ☐ No
Everyone wearing PPE: ☐ Yes ☐ No	Checklist Complete: ☐ Yes ☐ No

TODAY'S DATE	JOB/PROJECT INFORMATION
INITIALS	

HIGH TEMP	
LOW TEMP	

SCHEDULE INFORMATION	INSPECTION
Date of projected completion: / /	
Is project on schedule: ☐ Yes ☐ No	
Days behind:	

JOBSITE NOTES

INJURIES

Injuries on the job: ☐ Yes ☐ No	
If yes, was OSHA notified: ☐ Yes ☐ No	
Type of injury: ☐ First Aid ☐ Hospital	
(*Details MUST be listed under comments*)	

SAFETY

Toolbox topic:	Signage posted: ☐ Yes ☐ No
Everyone wearing PPE: ☐ Yes ☐ No	Checklist Complete: ☐ Yes ☐ No

TODAY'S DATE	JOB/PROJECT INFORMATION
INITIALS	

HIGH TEMP									
LOW TEMP									

SCHEDULE INFORMATION	INSPECTION
Date of projected completion: / /	
Is project on schedule: ☐ Yes ☐ No	
Days behind:	

JOBSITE NOTES

INJURIES	SAFETY	
Injuries on the job: ☐ Yes ☐ No	Toolbox topic:	Signage posted: ☐ Yes ☐ No
If yes, was OSHA notified: ☐ Yes ☐ No		
Type of injury: ☐ First Aid ☐ Hospital	Everyone wearing PPE: ☐ Yes ☐ No	Checklist Complete: ☐ Yes ☐ No
(Details MUST be listed under comments)		

TODAY'S DATE	JOB/PROJECT INFORMATION
INITIALS	

HIGH TEMP LOW TEMP									

SCHEDULE INFORMATION	INSPECTION
Date of projected completion: / /	
Is project on schedule: ☐ Yes ☐ No	
Days behind:	

JOBSITE NOTES

INJURIES

Injuries on the job: ☐ Yes ☐ No

If yes, was OSHA notified: ☐ Yes ☐ No

Type of injury: ☐ First Aid ☐ Hospital
(*Details MUST be listed under comments*)

SAFETY

Toolbox topic:	Signage posted: ☐ Yes ☐ No
Everyone wearing PPE: ☐ Yes ☐ No	Checklist Complete: ☐ Yes ☐ No

TODAY'S DATE	JOB/PROJECT INFORMATION
INITIALS	

HIGH TEMP									
LOW TEMP									

SCHEDULE INFORMATION	INSPECTION
Date of projected completion: / /	
Is project on schedule: ☐ Yes ☐ No	
Days behind:	

JOBSITE NOTES

(blank lined area)

INJURIES
Injuries on the job: ☐ Yes   ☐ No
If yes, was OSHA notified: ☐ Yes   ☐ No
Type of injury: ☐ First Aid   ☐ Hospital
(Details MUST be listed under comments)

SAFETY	
Toolbox topic:	Signage posted: ☐ Yes ☐ No
Everyone wearing PPE: ☐ Yes ☐ No	Checklist Complete: ☐ Yes ☐ No

TODAY'S DATE	JOB/PROJECT INFORMATION
INITIALS	

HIGH TEMP								
LOW TEMP								

SCHEDULE INFORMATION	INSPECTION
Date of projected completion: / /	
Is project on schedule: ☐ Yes ☐ No	
Days behind:	

JOBSITE NOTES

INJURIES	SAFETY	
Injuries on the job: ☐ Yes ☐ No	Toolbox topic:	Signage posted: ☐ Yes ☐ No
If yes, was OSHA notified: ☐ Yes ☐ No		
Type of injury: ☐ First Aid ☐ Hospital	Everyone wearing PPE: ☐ Yes ☐ No	Checklist Complete: ☐ Yes ☐ No
(Details MUST be listed under comments)		

TODAY'S DATE	JOB/PROJECT INFORMATION
INITIALS	

HIGH TEMP LOW TEMP									

SCHEDULE INFORMATION	INSPECTION
Date of projected completion: / /	
Is project on schedule: ☐ Yes ☐ No	
Days behind:	

JOBSITE NOTES

_____

_____

_____

_____

_____

_____

_____

_____

_____

_____

_____

_____

_____

_____

_____

_____

_____

INJURIES		SAFETY	
Injuries on the job: ☐ Yes ☐ No		Toolbox topic:	Signage posted: ☐ Yes ☐ No
If yes, was OSHA notified: ☐ Yes ☐ No			
Type of injury: ☐ First Aid ☐ Hospital (*Details MUST be listed under comments*)		Everyone wearing PPE: ☐ Yes ☐ No	Checklist Complete: ☐ Yes ☐ No

EQUIPMENT AND MATERIALS: PURCHASES AND/OR RENTALS			
DATE	INVOICE/JOB NUMBER	COMMENT	AMOUNT

It is important to document on a daily basis a brief description of the job status (progress and/or delays), safety issues/concerns, disagreements (owners, subcontractors, inspectors, etc.), issues with material deliveries and sales people, and anything that effects any aspect of your job or project. Remember, documentation is your best defense.

Download this form at www.DEWALT.com/guides

EQUIPMENT AND MATERIALS: PURCHASES AND/OR RENTALS			
DATE	INVOICE/JOB NUMBER	COMMENT	AMOUNT

EQUIPMENT AND MATERIALS: PURCHASES AND/OR RENTALS			
DATE	INVOICE/JOB NUMBER	COMMENT	AMOUNT

PURCHASE LOG

Download this form at www.DEWALT.com/guides

EQUIPMENT AND MATERIALS: PURCHASES AND/OR RENTALS			
DATE	INVOICE/JOB NUMBER	COMMENT	AMOUNT

EQUIPMENT AND MATERIALS: PURCHASES AND/OR RENTALS			
DATE	INVOICE/JOB NUMBER	COMMENT	AMOUNT

PURCHASE LOG

EQUIPMENT AND MATERIALS: PURCHASES AND/OR RENTALS			
DATE	INVOICE/JOB NUMBER	COMMENT	AMOUNT

Download this form at www.DeWALT.com/guides

EQUIPMENT AND MATERIALS: PURCHASES AND/OR RENTALS			
DATE	INVOICE/JOB NUMBER	COMMENT	AMOUNT

PURCHASE LOG

EQUIPMENT AND MATERIALS: PURCHASES AND/OR RENTALS			
DATE	INVOICE/JOB NUMBER	COMMENT	AMOUNT

Download this form at www.DEWALT.com/guides

EQUIPMENT AND MATERIALS: PURCHASES AND/OR RENTALS			
DATE	INVOICE/JOB NUMBER	COMMENT	AMOUNT

PURCHASE LOG

Download this form at www.DEWALT.com/guides

EQUIPMENT AND MATERIALS: PURCHASES AND/OR RENTALS			
DATE	INVOICE/JOB NUMBER	COMMENT	AMOUNT

Monthly Expense Report

MONTH _____

NAME	DATE
JOB(S)	

	WEEK1	WEEK2	WEEK3	WEEK4	WEEK5
DATE					
MILEAGE REIMBURSEMENT					
MEALS					
GAS					
LODGING					
WEEKLY TOTAL					

MONTHLY TOTAL _____

SUPERVISOR _____ DATE _____

Download this form at www.DeWALT.com/guides

Monthly Expense Report

MONTH _____

NAME	DATE
JOB(S)	

DATE	WEEK1	WEEK2	WEEK3	WEEK4	WEEK5
MILEAGE REIMBURSEMENT					
MEALS					
GAS					
LODGING					
WEEKLY TOTAL					

MONTHLY TOTAL _____

SUPERVISOR _____ DATE _____

Monthly Expense Report

MONTH _____

NAME	DATE
JOB(S)	

DATE	WEEK1	WEEK2	WEEK3	WEEK4	WEEK5
MILEAGE REIMBURSEMENT					
MEALS					
GAS					
LODGING					
WEEKLY TOTAL					

MONTHLY TOTAL _____

SUPERVISOR _____   DATE _____

Download this form at www.DeWALT.com/guides

EXPENSE REPORT

Monthly Expense Report

MONTH _____

NAME		DATE	
JOB(S)			

DATE	WEEK1	WEEK2	WEEK3	WEEK4	WEEK5
MILEAGE REIMBURSEMENT					
MEALS					
GAS					
LODGING					
WEEKLY TOTAL					

MONTHLY TOTAL _____

SUPERVISOR _____ DATE _____

Monthly Expense Report

MONTH _____

NAME	DATE
JOB(S)	

	WEEK1	WEEK2	WEEK3	WEEK4	WEEK5
DATE					
MILEAGE REIMBURSEMENT					
MEALS					
GAS					
LODGING					
WEEKLY TOTAL					

MONTHLY TOTAL _____

SUPERVISOR _____ DATE_____

EXPENSE REPORT

Download this form at www.DeWALT.com/guides

Monthly Expense Report

MONTH _____

NAME		DATE	
JOB(S)			

	WEEK1	WEEK2	WEEK3	WEEK4	WEEK5
DATE					
MILEAGE REIMBURSEMENT					
MEALS					
GAS					
LODGING					
WEEKLY TOTAL					

MONTHLY TOTAL _____

SUPERVISOR _____   DATE _____

Monthly Expense Report

MONTH _____

NAME	DATE
JOB(S)	

	WEEK1	WEEK2	WEEK3	WEEK4	WEEK5
DATE					
MILEAGE REIMBURSEMENT					
MEALS					
GAS					
LODGING					
WEEKLY TOTAL					

MONTHLY TOTAL _____

SUPERVISOR _____ DATE _____

EXPENSE REPORT

Download this form at www.DEWALT.com/guides

Monthly Expense Report

MONTH _____

NAME		DATE	
JOB(S)			

	WEEK1	WEEK2	WEEK3	WEEK4	WEEK5
DATE					
MILEAGE REIMBURSEMENT					
MEALS					
GAS					
LODGING					
WEEKLY TOTAL					

MONTHLY TOTAL _____

SUPERVISOR _____ DATE _____

Monthly Expense Report

MONTH _____

NAME	DATE
JOB(S)	

	WEEK1	WEEK2	WEEK3	WEEK4	WEEK5
DATE					
MILEAGE REIMBURSEMENT					
MEALS					
GAS					
LODGING					
WEEKLY TOTAL					

MONTHLY TOTAL _____

SUPERVISOR _____ DATE _____

EXPENSE REPORT

Download this form at www.DEWALT.com/guides

Monthly Expense Report

MONTH _____

NAME		DATE
JOB(S)		

	WEEK1	WEEK2	WEEK3	WEEK4	WEEK5
DATE					
MILEAGE REIMBURSEMENT					
MEALS					
GAS					
LODGING					
WEEKLY TOTAL					

MONTHLY TOTAL _____

SUPERVISOR _____ DATE _____

Monthly Expense Report

MONTH _____

NAME		DATE	
JOB(S)			

	WEEK1	WEEK2	WEEK3	WEEK4	WEEK5
DATE					
MILEAGE REIMBURSEMENT					
MEALS					
GAS					
LODGING					
WEEKLY TOTAL					

MONTHLY TOTAL _____

SUPERVISOR _____ DATE _____

EXPENSE REPORT

Download this form at www.DEWALT.com/guides

Monthly Expense Report

MONTH _____

NAME		DATE	
JOB(S)			

	WEEK1	WEEK2	WEEK3	WEEK4	WEEK5
DATE					
MILEAGE REIMBURSEMENT					
MEALS					
GAS					
LODGING					
WEEKLY TOTAL					

MONTHLY TOTAL _____

SUPERVISOR _____ DATE _____

VEHICLE LOG

DATE	MILEAGE	COMMENTS	OIL CHANGE	ROTATE / BALANCE TIRES	WHEEL ALIGNMENT	REPLACE TIRES	AIR FILTER	FUEL FILTER	WIPER BLADES	BRAKE SERVICE	BELTS AND HOSES	RADIATOR FLUSH & FILL	TRANSMISSION MAINT.	SPARK PLUGS	BATTERY	OTHER

Download this form at www.DEWALT.com/guides

Vehicle maintenance log form.

VEHICLE DESCRIPTION / YEAR, MAKE, AND MODEL / VIN NO.	DATE	MILEAGE	COMMENTS	OIL CHANGE	ROTATE / BALANCE TIRES	WHEEL ALIGNMENT	REPLACE TIRES	AIR FILTER	FUEL FILTER	WIPER BLADES	BRAKE SERVICE	BELTS AND HOSES	RADIATOR FLUSH & FILL	TRANSMISSION MAINT.	SPARK PLUGS	BATTERY	OTHER

VEHICLE LOG

DATE	MILEAGE	COMMENTS	OIL CHANGE	ROTATE / BALANCE TIRES	WHEEL ALIGNMENT	REPLACE TIRES	AIR FILTER	FUEL FILTER	WIPER BLADES	BRAKE SERVICE	BELTS AND HOSES	RADIATOR FLUSH & FILL	TRANSMISSION MAINT.	SPARK PLUGS	BATTERY	OTHER

Download this form at www.DeWALT.com/guides

VEHICLE DESCRIPTION	YEAR, MAKE, AND MODEL	VIN NO.															
DATE	MILEAGE	COMMENTS	OIL CHANGE	ROTATE / BALANCE TIRES	WHEEL ALIGNMENT	REPLACE TIRES	AIR FILTER	FUEL FILTER	WIPER BLADES	BRAKE SERVICE	BELTS AND HOSES	RADIATOR FLUSH & FILL	TRANSMISSION MAINT.	SPARK PLUGS	BATTERY	OTHER

Download this form at www.DeWALT.com/guides

DATE	MILEAGE	COMMENTS	OIL CHANGE	ROTATE / BALANCE TIRES	WHEEL ALIGNMENT	REPLACE TIRES	AIR FILTER	FUEL FILTER	WIPER BLADES	BRAKE SERVICE	BELTS AND HOSES	RADIATOR FLUSH & FILL	TRANSMISSION MAINT.	SPARK PLUGS	BATTERY	OTHER

DATE	MILEAGE	COMMENTS	OIL CHANGE	ROTATE / BALANCE TIRES	WHEEL ALIGNMENT	REPLACE TIRES	AIR FILTER	FUEL FILTER	WIPER BLADES	BRAKE SERVICE	BELTS AND HOSES	RADIATOR FLUSH & FILL	TRANSMISSION MAINT.	SPARK PLUGS	BATTERY	OTHER

VEHICLE DESCRIPTION

VIN NO.

YEAR, MAKE, AND MODEL

SAFETY CHECKLIST

General Requirements

☐ Injury log is up-to-date   ☐ OSHA Posters Displayed

☐ Emergency numbers posted   ☐ Safety Signs Posted

☐ First Aid Kit is readily available   ☐ First Aid kit contents checked

☐ Jobsite is properly illuminated   ☐ Eye washing facility available

☐ Temporary toilets available   ☐ Fire Extinguisher available

☐ Emergency Action Plan posted   ☐ MSDS available

Fall Protection and Protection from Falling Objects

☐ Each walking/working surface located 6' or more above a lower level is properly protected by guardrails or personal fall arrest systems.

☐ Height of required guardrails meet OSHA requirements (42" +/-3")

☐ Workers on walking/working surface are protected from tripping in or through holes (including skylights) by covers.

☐ Areas below construction zones have adequate protection from falling objects.

Jobsite Storage

☐ All materials are properly stacked, racked, blocked, or interlocked properly in a manner to prevent sliding, falling, or collapse.

☐ Maximum floor safe loads for stored items during construction are posted and not exceed.

☐ Passageways and aisles are clear to allow safe movement of equipment and workers.

☐ Materials stored inside structures under construction have not been placed within 6' of any hoistway or inside floor opening.

☐ Materials stored inside structures under construction have not been placed within 10' of an exterior wall that does not extend above the stacked material.

☐ All bagged materials are properly stacked in a manner that they are stepped back or cross-keyed at least every 10 bags high.

☐ Other than those needed for immediate consumption, materials are not stored on scaffolds or runways.

☐ Bricks are not stacked more than 7' – loose brick that are stacked have been properly tapered (2" back) each foot over 4'

☐ Masonry blocks are properly stacked (Those stacked higher than 6' are to be tapered back ½ block per tier above the 6')

☐ All nails have been removed from discarded lumber.

☐ Lumber is stacked, stable and self-supporting, on level and solidly supported sills.

- ☐ Lumber piles do not exceed 20'.
- ☐ Lumber to be handled manually does not exceed 16'.
- ☐ Structural steel, poles, pipe, bar stock and similar cylindrical materials that are not racked, have been properly stacked and blocked so as to prevent spreading or tilting.
- ☐ All storage areas are free from accumulated materials that constitute hazards from tripping, fire, explosion, or pest harborage.

Personal Protective Equipment

- ☐ Hard hats are being worn in areas where there is a potential for objects falling from above, bumps to head from fixed objects, or of accidental head contact with electrical hazards.
- ☐ Hearing protection, earplugs or earmuffs, is in place for workers exposed to high noise areas where sound exceeds acceptable levels. (See Chart on last page)
- ☐ Safety glasses or face shields are being worn in areas where operations can cause eye exposure to foreign objects.
- ☐ Work shoes or boots properly suited for work conditions are being worn.
- ☐ Steel-toed footwear is being worn by workers that are exposed to heavy equipment and/or falling objects.
- ☐ Proper gloves are being worn for hand protection. Gloves should be fit snugly.

Scaffolds:

- ☐ Fall prevention/protection is in place with Guardrails or Personal Fall Arrest Systems for any platform 10' or higher.
- ☐ Guardrails, midrails and toeboards are properly installed.
- ☐ All Planks are overlapped on a support at least 6" but not more than 12".
- ☐ Platform is tightly planked with proper material.
- ☐ Legs, posts, frames, poles, and uprights are situated on base plates and mud sills, or on a firm foundation.
- ☐ Legs, posts, frames, poles and uprights are plumb and braced.
- ☐ Properly braced with guy wires or attached to structure.
- ☐ Use has been verified to comply with Maximum intended load.
- ☐ Not placed closer than 10' from power lines.
- ☐ Workers below are protected from falling objects (area is barricaded or toeboards prevent objects from falling from above).

Ladders

- ☐ Visual inspection reveals no damage.
- ☐ Placed on level surface or secured (top and bottom) to prevent displacement.
- ☐ Not placed on unstable base to obtain additional height.

- ☐ Properly angled at a 4 to 1 ratio. (Self-supporting other than job-made) Example: 20' of working ladder length must be placed such that the base is 5' from the edge of the structure.
- ☐ Properly extends 3' above the point of support.
- ☐ Not placed in any location where it can be displaced by other work activities.
- ☐ If placed in a location where it CAN be displaced by other work activities, it is properly secured or a barricade has been erected to keep traffic away from the ladder.
- ☐ All locks on an extension ladder are properly engaged.
- ☐ Load rating supports all personnel, tools and equipment.

Temporary Stairs

- ☐ A stairway or ladder has been installed/placed in all locations where there is a break in elevation of 19" or more and no ramp, runway, embankment or personnel hoist.
- ☐ A 30" x 22" (Minimum) landing is in place at every 12' or less of vertical rise for all stairs.
- ☐ Handrails not less than 36" in height or properly installed on both sides of stairways that have 4 or more risers or more than 30" in height (whichever is less).

Trenching

- ☐ Protective systems are in place for trenches that are 5' or more in depth.
- ☐ Trenches that are 20' in depth or greater have been inspected by a registered professional engineer and a protective system has been installed according to his specifications.
- ☐ All trenches 5' or more in depth provide a means of egress such that a worker has no more than 25' of lateral travel to reach such exit.
- ☐ Spoils are not closer than 2' from open trench.
- ☐ Slope is proper according to soil type. (See chart on last page)

Disposal of Waste (1926.252)

- ☐ Debris is being removed at regular intervals during the course of construction
- ☐ An enclosed chute is provided for materials to be dropped more than 20' to any point outside the exterior walls of a building.
- ☐ A 42" tall barricade positioned at least 6' from the opening is installed when debris is being dropped through a floor without a chute. (Warning signs have been posted at each level)
- ☐ All scrap lumber, waste material and rubbish is being removed from the work area as debris is dropped (after debris handling ceases).
- ☐ All solvent waste, oily rags, and flammable liquids are properly discarded in fire resistant covered containers.

SAFETY CHECKLIST

Masonry

☐ Limited access has been established in areas where masonry walls are being constructed and is adequately restricted to entry by employees actively engaged in constructing the wall.

☐ Limited access zone is equal to the height of the wall to be constructed plus 4' and runs the entire length of the wall and has been established on the side of the wall which is to be unscaffolded.

☐ All masonry walls over 8' in height have been adequately braced to prevent overturning and collapse.

Structural Steel

☐ Permanent floors are being properly installed as the erection of the structural steel members progress and there is not more than eight stories between the erection floor and uppermost permanent floor.

☐ There is no more than four floors or 48' (whichever is less) of unfinished bolting or welding above the foundation or uppermost permanently secured floor.

☐ A fully planked/decked floor or nets have been installed to within two stories or 30' (whichever is less), directly under any erection work being performed.

Work Zone Traffic Safety

☐ Traffic control plans have been established in a manner that drivers, workers on foot and pedestrians can see and understand clearly the marked route.

☐ Flaggers are wearing high visibility clothing with a fluorescent background.

☐ Proper/required signage has been posted warning drivers that there will be flaggers ahead.

Confined Spaces

☐ All workers who enter confined spaces have been properly trained.

☐ Procedures have been established for confined space working and anyone entering such space has reviewed and understands the policy.

☐ Oxygen content (Breathable air) has been tested.

☐ A means of communication has been established for communication between a trained attendant and all personnel entering the confined space.

REFERENCE SECTION CONTENTS

Construction Math

    Rules to Avoid Costly Mistakes 1-1
    Converting from Inches to Feet 1-1
    Using a Measuring Tape. 1-1
    Rounding Numbers 1-1
    Profit-Markup vs. Gross Profit 1-2
    Pythagorean Theorem 1-2
    Identify "Margin" to maintain your profit goal. 1-2
    Multipliers 1-3
    3-4-5 Squaring System 1-3
    Rafter & Valley Calculations 1-3
    Roof Area from Plans 1-3
    Diagonal Squaring 1-3
    Coverage Area per Roll of Felt 1-3
    Waste Factors. 1-3
    Common Materials 1-3
    Gable Area 1-4
    Roof Area from Walking it. 1-4
    Required Slope for Excavations. 1-4
    Quantity of Studs 1-4
    Excavations 1-4
    Concrete by Cubic Yards 1-5
    Area Based on Geometry 1-5
    Calculating Volume for Common Geometric Shapes. 1-5
    Circumference of a Circle. 1-5
    Calculating Brick 1-6
    Mortar Required for Installing Brick 1-6
    Block Multiplier. 1-6
    Quantity of Black 1-6
    Mortar Volume Ratio. 1-6
    Concrete for Core Filling. 1-6
    Mortar. 1-6
    Drywall 1-7
    Estimating Tile 1-7
    Estimating Wall Paint 1-7
    Estimating Trim Paint 1-7

Standard Hand Signals for Controlling Crane Operations 1-8
Commonly Used Conversion Factors. 1-9
Commonly Used Geometrical Relationship 1-11
Decimal Equivalents of Fractions 1-12
Conversion Table for Temperature 1-13
Common Engineering Units and Their Relationship. 1-14
Metric Information. 1-15
    Base Units 1-15
    Supplementary SI Units 1-15
    Derived Metric Units with Compound Names 1-15
    SI Prefixes. 1-15
Metric Unit to Imperial Unit Conversion Factors. 1-16
Imperial Unit to Metric Unit Conversion Factors. 1-16
Abbreviations Used on Architectural Drawings 1-17

Graphic Symbols Used on Architectural Drawings. 1-22
    Standard Line Symbols 1-22
    Contours. 1-22
    Dimensions and Notes 1-22
    Identification Symbols 1-22
Symbols for Materials in Sections 1-23
Symbols for Materials in Elevation 1-25
Symbols for Walls in Section 1-26
Landscape Symbols 1-26
Door Symbols. 1-27
    Interior Doors Plan View 1-27
    Exterior Doors Plan View 1-27
Window Symbols 1-28
Typical Openings in Various Wall Constructions. 1-29
    Wood or Steel Stud Exterior Wall 1-29
    Brick Veneer Exterior Wall. 1-29
    Solid Masonry Exterior Wall 1-29
Receptacle Outlet Symbols 1-29
Lighting Outlet Symbols 1-30
Switch Symbols 1-30
Signaling and Communications Symbols. 1-30
Wire Symbols 1-30
Miscellaneous Electric Symbols 1-31
HVAC Symbols 1-31
    Register Symbols 1-31
    Ductwork Symbols 1-31
    Diffuser Symbols. 1-31
    Other Hvac Symbols. 1-31
Hot Water System Symbols 1-31
Air Conditioning Piping Symbols 1-32
Heating Piping Symbols 1-32
Plumbing Piping Symbols. 1-32
    Waste and Vents. 1-32
    Fire Piping. 1-32
    Water 1-32
    Other Piping 1-32
Plumbing Symbols 1-32
    Valves 1-32
    Fittings 1-33
    Other Symbols 1-33
Symbols Indicating Type of Pipe Material. 1-34
Symbols Indicating Materials Carried by Pipe 1-34
Symbols Indicating Flanged, Screwed, Welded, or Soldered Fittings 1-34
Fire Protection System Symbols. 1-35
Waterproofing Membranes and Coatings. 1-35
Waterproofing Materials in General Use. 1-35
    Built-Up Membranes 1-35
Waterproof Tips. 1-36
Weights of Construction Materials 1-37
    Brick and Masonry 1-37
    Concrete. 1-37
    Glass. 1-37
    Soft Woods 1-37
    Hardwoods 1-37

Construction Math

Rules to avoid costly mistakes

1. **ALWAYS USE A CALCULATOR**
   NEVER add anything in your mind. And you should always perform each complete calculation at least twice. Onin to determine if you get the same answer. If you don't, try it again.

2. **WRITE IT OUT**
   If you don't write it down, how are you going to check it? Professionals who use math on a daily basis: accountants, engineers, and the like, are sometimes the most difficult students that we encounter when it comes to math. Why? Because they refuse to write it down and they will not use a calculator for what they feel is "simple math."

3. **ALWAYS MULTIPLY "LIKE NUMBERS".**
   For example: do not multiply 10' by 4" – you must either convert the 10' to inches or the 4" to feet. This is a common mistake in doing material take-offs.

4. **NEVER ROUND UNTIL YOU GET TO THE END**
   For example: if you are calculating bundles of shingles you begin by computing squares. If you arrive at 15.4 squares, DO NOT round to 16. Multiply it out first.

5. **DRAW IT OUT**
   If you have complex calculations, it always helps to get graph paper and plot it out.

Converting from Inches to Feet

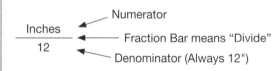

Inches — Numerator

Fraction Bar means "Divide"

12 — Denominator (Always 12")

Inches to Feet	
1"	.083
2"	.17
3"	.25
4"	.33
5"	.42
6"	.50
7"	.58
8"	.67
9"	.75
10"	.83
11"	.92
12"	1.00

Step 1
Place the number (Inches) you want to convert to feet over 12.

Step 2
Divide the Numerator by the Denominator

Step 3
Round to the nearest 100th (unless otherwise specified)

The first and possibly the most important steps in doing an estimate is to make sure you are multiplying "like" numbers. Never multiply inches by feet.

Using a Measuring Tape

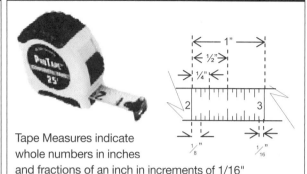

Tape Measures indicate whole numbers in inches and fractions of an inch in increments of 1/16"

To use a measuring tape, you will need to understand how to add and subtract fractions.

Step 1 —— The first to adding or subtracting fractions is to reduce the fractions to their lowest possible terms.

Step 2 —— Next, find the lowest common denominator for each.

Step 3 —— To convert to the lowest common denominator, find the lowest common multiple. This is the lowest number that will divide into each.

Example: $2/8 + 3/16 = $ __?__

Step 1:   Reduce 2/8 to 1/4; 3/16 does not reduce.

Step 2:   The lowest common denominator is 16 1/4 will become 4/16 and 3/16 will remain the same.

Step 3:   4/16 + 3/16 = 7/16.

Rounding Numbers

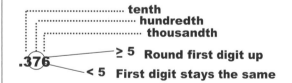

tenth
hundredth
thousandth

.376

≥ 5 **Round first digit up**

< 5 **First digit stays the same**

Identify the number to the right of the "place" you intend to round. (for example, if you are rounding to the nearest 100th place, identify the third number to the right of the decimal) If the digit to the right of the place you are rounding to is 5 or greater, increase the "place" you are rounding to by 1 and drop all numbers to the right of it. If the number is less than five, do nothing but drop all numbers to the right of the place you are rounding to.

REFERENCE

Construction Math

Profit – Markup vs. Gross Profit

Markup and Gross Profit are often confused. Markup is simply the factor you apply to your estimated job costs to determine what you will charge your customer. For example: You estimate the cost to do a job is $100,000 and you want to "markup" the job 30%. You determine the contract you will sign with the customer will be for $130,000 (100,000 × 1.30).

Did you really make 30%? The answer is "no." Gross profit is computed by subtracting the cost of sales (Materials, labor etc) from your contract price. The "profit" percentage is equal to gross profit divided by the contract price.

$$\frac{30,000}{130,000} \times .23 \text{ (or 23\%)}$$

This is your "percentage" of profit.

Understanding this concept is critical because you should be adding a percentage to "costs" that will cover overhead and profit. Your profit should be over and above your salary. If you determine your Project Overhead runs about 10%, your Company Overhead runs about 12% and you always want to make a 10% profit, you should be adding 32% to all estimates. To do this properly, subtract 32% (what you need to make over your costs) from 100% (which represents the bid amount. 100 − 32 = 68, you should divide the cost to complete a job by this percent. For example, if a job "cost" totaled $80,000, you should divide 80,000 by .68 to determine your bid amount should be: $117,647.05 (this will give you 32% above your "costs")

Identify "Margin" to maintain your profit goal.

Step 1.
Establish a Profit – decide how much you want to put in your pocket (Profit) after you pay yourself a fair market salary. For our example – we will use 10%.

Step 2.
Identify Overhead Expenses – go through last year's bills and calculate what it costs to run a job (this is your Project Overhead). Next, look at what it costs to operate your business. What does it cost to run an office and pay your managers, secretaries, bookkeepers etc. Convert this to a percentage of what your revenue was for that particular year. For example: if your overhead was $120,000 and your revenue was $800,000, your overhead represents 15% of your revenue.

Step 3
Compute projected gross profit – 10% Profit + 15% Overhead = 25% Gross Profit. 100 − 25% = 75% - you should divide all project "costs" by .75 to determine your estimate.

Example: You are preparing a bid for a remodeling job. The materials and labor will "cost" you $68,000. To accurately prepare the bid, divide 68,000 by .75 = $90,666. This is what your contract amount should be.

Pythagorean Theorem

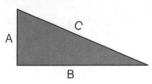

$$C = \sqrt{A^2 + B^2}$$

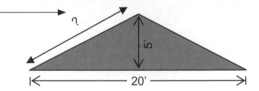

If you know two of the three measurements, you can find the third. In construction, "C" (Hypotenuse) is typically what we are seeking to learn. "C" can be a rafter.

"C" can also be a measurement we need to know in order to confirm the corner is "Square" or a perfect 90 degree angle. This is commonly known in the field as the 3-4-5 system of squaring or Diagonal Squaring.

Step 1 —— Determine "B" is 10' (half of the 20' Span)

Step 2 —— Use the formula. $C = \sqrt{5^2 + 10^2}$

Step 3 $C = \sqrt{25 + 100}$ **Step 4** $C = \sqrt{125}$

The Hypotenuse (Rafter) is 11.18'

Multipliers

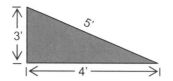

This indicates the roof is "Sloped" 5 in 12

If a roof is sloped 5 in 12, it rises 5" for each 12" of horizontal run.

Slopes for roofs are typically provided in a ratio of the number of units it rises per 12 units of run. (see above) Multipliers have been provided below based on each of the common slopes used in construction. The multiplier is the increased length of the hypotenuse for each foot of run. Simply multiply the run (Flat dimension) by the multiplier provided and avoid using the Pythagorean Theorem Formula.

	Run Multiplier	
Roof Slope	Rafter	Valley
2 in 12	1.014	1.424
3 in 12	1.031	1.436
4 in 12	1.054	1.453
5 in 12	1.083	1.474
6 in 12	1.118	1.500
7 in 12	1.158	1.530
8 in 12	1.202	1.564
9 in 12	1.357	1.601
10 in 12	1.302	1.642
11 in 12	1.357	1.685
12 in 12	1.413	1.732
13 in 12	1.474	1.782
14 in 12	1.537	1.833
15 in 12	1.601	1.887
16 in 12	1.667	1.944
17 in 12	1.734	2.002
18 in 12	1.803	2.032
19 in 12	1.875	2.123
20 in 12	1.948	2.186
21 in 12	2.010	2.250
22 in 12	2.083	2.315
23 in 12	2.167	2.382

3-4-5 Squaring System

Step 1
Measure 3' from one corner.

Step 2
From the same corner, measure 4' in the opposite direction.

Step 3
If the corner is 90 degrees, the diagonal measurement taken from the two points will be 5'. (use any increment of 3-4-5 such as 6-8-10)

Rafter & Valley Calculations

Step 1
Measure the run (Flat dimension)

Step 2
Identify the multiplier that corresponds with the roof slope. (Table to the left)

Step 3
Multiply the run by the multiplier.

Step 4
Round to the next "even" number.

Roof Area from Plans

Step 1
Add the overhang to dimensions.

Step 2
Multiply Length by Width.

Step 3
Multiply the "area" by the "Rafter" multiplier (to the left) that corresponds with the roof slope specified on your plans and divide by 100 for squares.

Diagonal Squaring

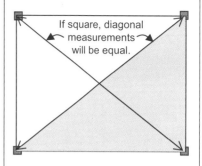

If square, diagonal measurements will be equal.

Pythagorean Theorem can easily be applied for rectangular layouts without using a calculator. Simply take the diagonal measurements – if they are equal, the building is square.

It is recommended that you use the 3-4-5 system on layouts with many offsets and to s =

Coverage area per roll of felt

15# Felt Underlayment.............400 SF

30# Felt Underlayment.............200 SF

Waste Factors

Simple roofs (Gable) .... 1 – 5%

Complex roofs .... 5 – 10%

(with multiple open valleys)

Common Materials

Shingle Type	Pkgs Per Sq	Dimensions	Nails Per Sq
3 Tab	3	36" × 12"	
4 Tab (Architectural)	3	36" × 12"	
Wood Shingles	5	Avg Width 4" (4" to weather)	2 lbs (3d)
Wood Shingles	4	Avg Width 4" (5" to weather)	3.2 lbs (3d)

REFERENCE

Gable Area

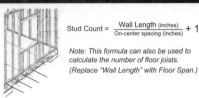

Square Feet

12
6

← Run — 15' →
← Span — 30' →

Step 1
Convert the span to run (in feet and decimals of feet)

Step 2
Identify the slope factor (such as 6:12)

Step 3
Multiply the run by the vertical rise and divide by 12 to convert to feet – this is the Gable Height.

Step 4
Multiply the Gable Height (in feet) by the run – this is the square feet of the Gable Area.

Quantity of Studs

$$\text{Stud Count} = \frac{\text{Wall Length (inches)}}{\text{On-center spacing (inches)}} + 1$$

Note: This formula can also be used to calculate the number of floor joists. (Replace "Wall Length" with Floor Span.)

Step 1
Measure the length of the wall.

Step 2
Make sure the length is in inches.

Step 3
Divide the length by the "on- center" spacing (in inches) and round to the next whole number (round up).

Step 4
Add one.

Example: Studs are to be installed 16" on center. The length of the wall is 14'6".
Step 1. Wall is 14.5'
Step 2. Convert into inches, 14.5 × 12 = 174"
Step 3. 174 / 16 = 10.875 (round to 11)
Step 4. 11 + 1 = 12

Note: Additional studs should be considered for corners and openings

Roof Area from Walking it

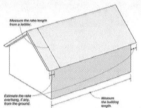

Step 1
Measure the Height from the bottom of the roof line to the ridge board.

Step 2
Measure the Length (include overhang)

Step 3
Divide the total area by 100 (this is the number of "squares" required)

Step 4
Multiply the squares by the number of bundles needed per square.

Excavations

Top of the excavation

Bottom of the excavation

$$\frac{\text{Average Length} \times \text{Average Width} \times \text{Average Depth}}{27}$$

Step 1
Identify Soil Type to determine the required slope.

Step 2
Multiply the Average Depth by the Slope Ratio (if ½:1, Depth × .5).

Step 3
Add the product of Step 2 to each side of the Length and each side of the width.

Step 4
Length at the top of the excavation + Length at the bottom of the excavation divided by 2 (to get Average).

Step 5
Width at the top of the excavation + Width at the bottom of the excavation divided by 2 (to get Average).

Step 6
Multiply the Average Length by the Average Width by the Average Depth to get Cubic Feet.(L × W × D)

Step 7
Divide Cubic Feet by 27 to get Cubic Yards.

Required Slope for Excavations

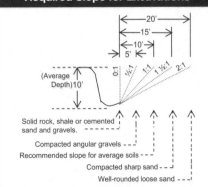

← 20' →
← 15' →
← 10' →
5'

(Average Depth)10'

0:1 ½:1 1:1 1½:1 2:1

Solid rock, shale or cemented sand and gravels.
Compacted angular gravels
Recommended slope for average soils
Compacted sharp sand
Well-rounded loose sand

Concrete by Cubic Yards

Cubic Feet = L X W X D

Cubic Yards = $\dfrac{L \times W \times D}{27}$

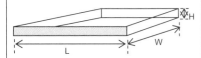

Step 1
Multiply the Length by the width to calculate area.

Step 2
Multiply the area by the depth to calculate cubic feet. (Make sure the depth has been converted from inches to feet)

Step 3
Most concrete is ordered by the "yard" (or cubic yard) so you must now divide the cubic feet by 27.

Table for estimating Concrete by Cubic Yard

Depth	CY	Depth	CY	Depth	CY	Depth	CY	Depth	CY
2"	162	3.25"	99.7	4.5"	72				
2.25"	144	3.5"	92.6	4.75"	68.2				
2.5"	129.6	3.75"	86.4						
2.75"	117.8	4"	81						
3"	108	4.25"	76.2						

Divide the SF of your slab by the number corresponding with the depth of your pour to determine the number of cubic yards.

Calculating Volume for Common Geometric Shapes

Cylinder
Volume = 0.7854^2 x A x B

Example: Calculate Cubic Feet (Volume) in a tank that is 20' tall with an 8' diameter.

$0.7854 \times 8 \times 8 \times 20 = 1005.312$
(to convert to cubic yards, divide by 27)

Pyramid
Volume = $\dfrac{A \times B \times C}{3}$

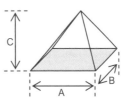

Example: Calculate Cubic Feet (Volume) in a pyramid with dimensions of 10' × 10' with a 14' height.

$10 \times 10 \times 14 = 1400$
1400 divided by 3 = 466.67
(to convert to cubic yards, divide by 27)

Cone
Volume = 1.0472^2 x B x C

Example: Calculate Cubic Feet (Volume) in a cone with 5' radius that is 14' tall.

$1.0472 \times 5 \times 5 \times 14 =$
(to convert to cubic yards, divide by 27)

Area based on Geometry

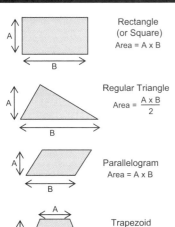

Rectangle (or Square)
Area = A x B

Regular Triangle
Area = $\dfrac{A \times B}{2}$

Parallelogram
Area = A x B

Trapezoid
Area = $\left(\dfrac{A + B}{2}\right) \times C$

Right Triangle
Area = $\dfrac{A \times B}{2}$

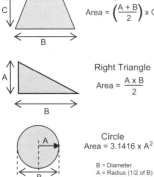

Circle
Area = $3.1416 \times A^2$

B = Diameter
A = Radius (1/2 of B)

Circumference of a Circle

Circumference is Linear Distance

Circumference = B x 3.1416

B = Diameter
A = Radius (1/2 of B)

Example: Calculate the Linear Feet around a circle that has a 12' diameter.
$12 \times 3.1416 = 37.6992'$

EXAMPLE
A painter charges by the Square foot and needs to determine the Area of a Column that is 25' tall and 5' in diameter. He will charge 1.25 per square foot – there are 10 columns.

Step 1
Calculate the circumference of the column.
$5 \times 3.1416 = 15.708$

Step 2
Multiply the Circumference by the Height of the column.
$15.708 \times 25 = 392.7$ SF

Step 3
Multiply the SF by $1.25
$392.7 \times \$1.25 = \$490.88$

Step 4
Multiply the cost to paint one column by 10 (number of columns)
$\$490.88 \times 10 = \$4908.80$

REFERENCE

Calculating Brick

Brick Multiplier	Brick Quantity
$$\frac{144}{H^* \times L^* \text{ of brick}}$$ *Must add mortar joint first*	Area* × Multiplier *Deduct for Openings*

Step 1
Add one side of the brick mortar joint to face dimensions of the brick.

Step 2
Multiply the Height of the brick by the Length of the brick to determine the face are in square inches. (or use the table)

Step 3
Divide 144 Square inches by the face area of the brick. (this is the multiplier – how many bricks you will need per square foot.)

Step 4
Calculate the area of the space for installing brick – always deduct the square feet of the openings.

Step 5
Multiply the area by the brick multiplier (can also be found in the table below) and add 5 – 10% for waste.

Type	Brick Size			Mortar Joints				
	Height	Length	Width	¼"	3/8"	½"	5/8"	¾"
Common	2 ¼"	8"	3 ¾"	7	6.6	6.2	5.8	5.5
Large St	1 5/8"	15 ¾"	3 ¾"	4.8	4.5	4.2	3.9	3.3
Small St	1 ¼"	8"	3 ¾"	5.3	5.1	4.8	4.5	4.3
Jumbo	2 1/8"	11 ½"	5 ½"	1.16	1.05	9.7	4.1	3.9
Norman	2 ¾"	8 ¾"	4"	5.2	4.9	4.6	4.3	4.1
Roman	2 ¼"	11 ½"	3 ¾"	4.9	4.6	4.4	4.1	3.9

Identify the size of your mortar joint, follow the table down to the number that corresponds with your brick size. Multiply this number by the area for which you will be installing brick. The product is the number of brick you will need.

Block Multiplier

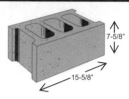

Step 1
Add 3/8" Mortar joint to block dimensions = 16" × 8".

Step 2
Multiply 16 × 8 = 128" (this is the face area of the block in square inches)

Step 3
144 divided by 128 = 1.125 (144 is the square inches in a square foot – you will need 1.125 block per square foot)

Quantity of Block

Step 1
Add the total Linear Feet of walls to be blocked.

Step 2
Multiply the total Linear Feet by the Height of the building to determine the Area to be blocked.

Step 3
Subtract the total square feet of all openings from the Area.

Step 4
Multiply the Area by 1.125.

Example: Calculate the block needed for a wall that is 50' long, 10' high with 80 square feet of openings.

Step 1. 50'   Step 2. 50' × 10' = 500 SF

Step 3. 500 – 80 = 420   Step 4. 420 × 1.125 = 473

Determine the number of courses

Height (in feet) divided by .67

Mortar Volume Ratio

	Portland Mix		
Type	Portland Cement	Hydrated Lime	Sand
M	1	¼	3
S	1	½	4.5
N	1	1	6
O	1	2	9

Mortar Volume Ratio

	Masonry Mix		
Type	Portland Cement	Hydrated Lime	Sand
M	1	1	6
S	½	1	4.5
N	N/A	1	3
O	N/A	1	3

Mortar required for installing brick

CF of mortar required to lay 1,000 Brick

Joint Thickness in inches	Thickness of the wall in Inches					
	4"	8"	12"	16"	20"	24"
1/8"	2.9	5.6	6.5	7.1	7.3	7.5
1/4"	5.7	8.7	9.7	10.2	10.5	10.7
3/8"	8.7	11.8	12.9	13.4	13.7	14
1/2"	11.7	15	16.2	16.8	17.1	17.3
5/8"	14.8	18.3	19.5	20.1	20.5	20.7
3/4"	17.9	21.7	23	23.6	24	24.2
7/8"	21.1	25.1	26.5	27.1	27.5	27.8
1"	24.4	28.6	30.1	30.8	31.2	31.5

Step 1
Identify the joint thickness.

Step 2
Identify the thickness of your wall.

Step 3
Identify the multiplier that corresponds with Step 1 and Step 2.

Step 4
Divide the total number of brick you are installing by 1000.

Step 5
Use the multiplier from Step 3 to multiply by the number from Step 4. This is the Cubic Feet of mortar you will need.

Concrete for Core Filling

6 × 8 × 16	0.17 CF / Block
8 × 8 × 16	0.25 CF / Block
10 × 8 × 16	0.33 CF / Block
12 × 8 × 16	0.39 CF / Block
6 × 8 × 16 Bond Beam	0.173 CF per LF
8 × 8 × 16 Bond Beam	0.22 CF per LF
8 × 8 × 16 Deep Bond Beam	0.46 CF per LF
12 × 8 × 16 Bond Beam	0.37 CF per LF
12 × 8 × 16 Deep Bond Beam	0.74 CF per LF

Mortar

Block	3 bags per 100 block (9 CF Sand)
Face Brick Modular	7 bags per 1000 brick (21 CF Sand)
Oversized Brick	8 bags per 1000 brick
Utility Brick	10 bags per 1000 brick

Drywall

Step 1
Measure the length of each wall and add them together.

Step 2
Multiply the length times the ceiling height to determine the wall area.

Step 3
Multiply the length by the width of the room to determine the area of the ceiling.

Step 4
Add the Area of the Walls to the Area of the ceiling.

Step 5
Divide the total area by the area of the Board to be installed. (if 4 × 8 — divide by 32)

Note: Typically, you do not deduct for openings. Keep calculations separate for the ceiling if different thickness/size boards are being installed.

Ready Mix (LB)
Wall Area × 0.14

Tape (LF)
Wall Area × 0.48

Joint Compound (LB)
Wall Area × 0.09

Nails (LB)
Wall Area × 0.005

Screws (EA)
Wall Area × 1.25

Estimating Wall Paint

Step 1
Add the length of each wall together.

Step 2
Multiply the total length of the walls by the ceiling height (this is the wall area)

Step 3
Subtract the square feet of the openings from the total wall area.

Step 4
Divide the remaining square feet by 350* to determine the number of gallons needed and increase this number by the number of coats to be applied. (*Use the table below or see paint can)

Example: Applying two coats, how many gallons of paint will you need for a room that is 12 × 14 with 9' ceilings? (Note: Total SF of openings is 45)

Step 1: 12 + 12 + 14 + 14 = 52

Step 2: 52 × 9 = 468

Step 3: 468 − 45 = 423

Step 4: 423 / 350 = 1.21 × 2 coats

Coverage Area for Interior Paint

Surface and Paint Type	First Coat
Concrete Block	200
Gypsum Board (flat)	450
Gypsum Board (gloss)	400
Plaster (texture)	250
	SF per Gallon

Coverage Area for Exterior Paint

Surface and Paint Type	First Coat	Second Coat	Third Coat
Brick (Oil Base)	200	400	N/A
Brick (Water Base)	100	150	N/A
Concrete Floor (Paint)	450	600	600
Concrete Floor (Stain)	510	480	N/A
Gutter	200	N/A	N/A
Porch Floor (Wood)	378	540	576
Siding, shingle (Paint)	342	423	N/A
Siding, shingle (Stain)	150	225	N/A
Siding (Paint)	420	520	620
Stucco* (oil)	153	360	360
Stucco* (water/cement)	99	135	N/A
Trim	850	900	972
*Medium Texture		Square feet per Gallon	

Using the Table

1. Identify your surface in the corresponding "exterior" or "interior" table.

2. Divide your total area to be painted by the number in the first coat column.

3. Add the "second" coat. (if you are applying 2nd coat)

Estimating Trim Paint

Step 1
Add the total linear feet of trim..

Step 2
Multiply the total length by .5 (or exact width of the trim in feet)

Step 3
Divide by 350 to determine the number of gallons needed.

Estimating Tile

Step 1
Multiply the length times the width of the tile piece (in inches) to determine the Area in square inches.

Step 2
Divide 144 (the square inched in a square foot) by the area of the tile piece to be installed. The result is how many pieces you will need per square foot.

Step 3
Multiply the length by the width of the room to determine the Area in square feet that you will be installing tile.

Step 4
Multiply this Area by the multiplier from Step 2

Tiles per SF	
3" × 3"	16
4" × 4"	9
4 ¼" × 4 ¼"	8
6" × 6"	4
6" × 8"	3
6" × 12"	2
8" × 8"	2.25
9" × 9"	1.78
10" × 10"	1.45
12" × 12"	1
9" × 18"	.89

STANDARD HAND SIGNALS FOR CONTROLLING CRANE OPERATIONS

	HOIST Forearm vertical, forefinger pointing up, move hand in small, horizontal circles.		**LOWER THE BOOM AND RAISE THE LOAD** Arm extended, thumb pointing down, flex fingers in and out.
	LOWER Arm extended downward, forefinger pointing down, move hand in small, horizontal circles.		**SWING** Arm extended, point with finger in direction of swing.
	USE MAIN HOIST Tap fist on head; then use regular signals.		**STOP** Arm extended, palm down, hold.
	USE WHIPLINE Tap elbow with one hand; then use regular signals.		**EMERGENCY STOP** Arm extended, palm down, move hand rapidly right and left.
	RAISE BOOM Arm extended, fingers closed, thumb pointing upward.		**TRAVEL** Arm extended forward, hand open and slightly raised, pushing motion in direction of travel.
	LOWER BOOM Arm extended, fingers closed, thumb pointing down.		**EXTEND BOOM** Both fists in front of body with thumbs pointing outward.
	MOVE SLOWLY One hand gives motion signal, other hand motionless in front of hand giving the motion signal.		**RETRACT BOOM** Both fists in front of body with thumbs pointing toward each other.
	RAISE THE BOOM AND LOWER THE LOAD Arm extended, thumb pointing up, flex fingers in and out.		A crane operator should always move loads in accordance to the established code of signals by way of a signaler and are preferred and commonly used. If there is any question on your ability to see the signal man or the load, use radio communication or another signal person.

COMMONLY USED CONVERSION FACTORS

MULTIPLY	BY	TO OBTAIN
Acres	43,560	Square feet
Acres	1.562×10^{-3}	Square miles
Acre-Feet	43,560	Cubic feet
Amperes per sq. cm.	6.452	Amperes per sq. in.
Amperes per sq. in.	0.1550	Amperes per sq. cm.
Ampere-Turns	1.257	Gilberts
Ampere-Turns per cm.	2.540	Ampere-turns per in.
Ampere-Turns per in.	0.3937	Ampere-turns per cm.
Atmosheres	76.0	Cm. of mercury
Atmospheres	29.92	Inches of mercury
Atmospheres	33.90	Feet of water
Atmospheres	14.70	Pounds per sq. in.
British thermal units	252.0	Calories
British thermal units	778.2	Foot-pounds
British thermal units	3.960×10^{-4}	Horsepower-hours
British thermal units	0.2520	Kilogram-calories
British thermal units	107.6	Kilogram-meters
British thermal units	2.931×10^{-4}	Kilowatt-hours
British thermal units	1,055	Watt-seconds
BTU per hour	2.931×10^{-4}	Kilowatts
BTU per minute	2.359×10^{-2}	Horsepower
BTU per minute	1.759×10^{-2}	Kilowatts
Bushels	1.244	Cubic feet
Centimeters	0.3937	Inches
Circular mils	5.067×10^{-6}	Square centimeters
Circular mils	0.7854×10^{-6}	Square inches
Circular mils	0.7854	Square mils
Cords	128	Cubic feet
Cubic centimeters	6.102×10^{-6}	Cubic inches
Cubic feet	0.02832	Cubic meters
Cubic feet	7.481	Gallons
Cubic feet	28.32	Liters
Cubic inches	16.39	Cubic centimeters
Cubic meters	35.31	Cubic feet
Cubic meters	1.308	Cubic yards
Cubic yards	0.7646	Cubic meters
Degrees (angle)	0.01745	Radians
Dynes	2.248×10^{-6}	Pounds
Ergs	1	Dyne-centimeters
Ergs	7.37×10^{-6}	Foot-pounds
Ergs	$10-7$	Joules

MULTIPLY	BY	TO OBTAIN
Farads	106	Microfarads
Fathoms	6	Feet
Feet	30.48	Centimeters
Feet of water	.08826	Inches of mercury
Feet of water	304.8	Kg. per square meter
Feet of water	62.43	Pounds per square ft.
Feet of water	0.4335	Pounds per square in.
Foot-pounds	1.285×10^{-2}	British thermal units
Foot-pounds	5.050×10^{-7}	Horsepower-hours
Foot-pounds	1.356	Joules
Foot-pounds	0.1383	Kilogram-meters
Foot-pounds	3.766×10^{-7}	Kilowatt-hours
Gallons	0.1337	Cubic feet
Gallons	231	Cubic inches
Gallons	3.785×10^{-3}	Cubic meters
Gallons	3.785	Liters
Gallons per minute	2.228×10^{-3}	Cubic feet per sec.
Gausses	6.452	Lines per square in.
Gilberts	0.7958	Ampere-turns
Henries	103	Millihenries
Horsepower	42.41	BTU per min.
Horsepower	2,544	BTU per hour
Horsepower	550	Foot-pounds per sec.
Horsepower	33,000	Foot-pounds per min.
Horsepower	1.014	Horsepower (metric)
Horsepower	10.70	Kg. calories per min.
Horsepower	0.7457	Kilowatts
Horsepower (boiler)	33,520	B.t.u. per hour
Horsepower-hours	2,544	British thermal units
Horsepower-hours	1.98×106	Foot-pounds
Horsepower-hours	2.737×105	Kilogram-meters
Horsepower-hours	0.7457	Kilowatt-hours
Inches	2.540	Centimeters
Inches of mercury	1.133	Feet of water
Inches of mercury	70.73	Pounds per square ft.
Inches of mercury	0.4912	Pounds per square in.
Inches of water	25.40	Kg. per square meter
Inches of water	0.5781	Ounces per square in.
Inches of water	5.204	Pounds per square ft.
Joules	9.478×10^{-4}	British thermal units
Joules	0.2388	Calories

1-9

REFERENCE

MULTIPLY	BY	TO OBTAIN
Joules	107	Ergs
Joules	0.7376	Foot-pounds
Joules	2.778×10^{-7}	Kilowatt-hours
Joules	0.1020	Kilogram-meters
Joules	1	Watt-seconds
Kilograms	2.205	Pounds
Kilogram-calories	3.968	British thermal units
Kilogram meters	7.233	Foot-pounds
Kg per square meter	3.281×10^{-3}	Feet of water
Kg per square meter	0.2048	Pounds per square ft.
Kg per square meter	1.422×10^{-3}	Pounds per square in.
Kilolines	103	Maxwells
Kilometers	3.281	Feet
Kilometers	0.6214	Miles
Kilowatts	56.87	B.t.u. per min.
Kilowatts	737.6	Foot-pounds per sec.
Kilowatts	1.341	Horsepower
Kilowatts-hours	3409.5	British thermal units
Kilowatts-hours	2.655×10^6	Foot-pounds
Knots	1.152	Miles
Liters	0.03531	Cubic feet
Liters	61.02	Cubic inches
Liters	0.2642	Gallons
Log Ne or in N	0.4343	Log10 N
Log N	2.303	Loge N or in N
Lumens per square ft.	1	Footcandles
Maxwells	10-3	Kilolines
Megalines	106	Maxwells
Megaohms	106	Ohms
Meters	3.281	Feet
Meters	39.37	Inches
Meter-kilograms	7.233	Pound-feet
Microfarads	10-6	Farads
Microhms	10-6	Ohms
Microhms per cm. cube	0.3937	Microhms per in. cube
Microhms per cm. cube	6.015	Ohms per mil. foot
Miles	5,280	Feet
Miles	1.609	Kilometers
Miner's inches	1.5	Cubic feet per min.
Ohms	10-6	Megohms
Ohms	106	Microhms
Ohms per mil foot	0.1662	Microhms per cm. cube
Ohms per mil foot	0.06524	Microhms per in. cube

MULTIPLY	BY	TO OBTAIN
Poundals	0.03108	Pounds
Pounds	32.17	Poundals
Pound-feet	0.1383	Meter-Kilograms
Pounds of water	0.01602	Cubic feet
Pounds of water	0.1198	Gallons
Pounds per cubic foot	16.02	Kg. per cubic meter
Pounds per cubic foot	5.787×10^{-4}	Pounds per cubic in.
Pounds per cubic inch	27.68	Grams per cubic cm.
Pounds per cubic inch	2.768×10^{-4}	Kg. per cubic meter
Pounds per cubic inch	1.728	Pounds per cubic ft.
Pounds per square foot	0.01602	Feet of water
Pounds per square foot	4.882	Kg. per square meter
Pounds per square foot	6.944×10^{-3}	Pounds per sq. in.
Pounds per square inch	2.307	Feet of water
Pounds per square inch	2.036	Inches of mercury
Pounds per square inch	703.1	Kg. per square meter
Radians	57.30	Degrees
Square centimeters	1.973×10^5	Circular mils
Square Feet	2.296×10^{-5}	Acres
Square Feet	0.09290	Square meters
Square inches	1.273×10^6	Circular mils
Square inches	6.452	Square centimeters
Square Kilometers	0.3861	Square miles
Square meters	10.76	Square feet
Square miles	640	Acres
Square miles	2.590	Square kilometers
Square Millimeters	1.973×10^3	Circular mils
Square mils	1.273	Circular mils
Tons (long)	2,240	Pounds
Tons (metric)	2,205	Pounds
Tons (short)	2,000	Pounds
Watts	0.05686	B.t.u. per minute
Watts	107	Ergs per sec.
Watts	44.26	Foot-pounds per min.
Watts	1.341×10^{-3}	Horsepower
Watts	14.34	Calories per min.
Watts-hours	3.412	British thermal units
Watts-hours	2,655	Footpounds
Watts-hours	1.341×10^{-3}	Horsepower-hours
Watts-hours	0.8605	Kilogram-calories
Watts-hours	376.1	Kilogram-meters
Webers	108	Maxwells

COMMONLY USED GEOMETRICAL RELATIONSHIP	
Diameter of a circle × 3.1416	Circumference.
Radius of a circle × 6.283185	Circumference.
Square of the radius of a circle × 3.1416	Area.
Square of the diameter of a circle × 0.7854	Area.
Square of the circumference of a circle × 0.07958	Area.
Half the circumference of a circle × half its diameter	Area.
Circumference of a circle × 0.159155	Radius.
Square root of the area of a circle × 0.56419	Radius.
Circumference of a circle × 0.31831	Diameter.
Square root of the area of a circle × 1.12838	Diameter.
Diameter of a circle × 0.866	Side of an inscribed equilateral triangle.
Diameter of a circle × 0.7071	Side of an inscribed square.
Circumference of a circle × 0.225	Side of an inscribed square.
Circumference of a circle × 0.282	Side of an equal square.
Diameter of a circle × 0.8862	Side of an equal square.
Base of a triangle × one-half the altitude	Area.
Multiplying both diameters and .7854 together	Area of an ellipse.
Surface of a sphere × one-sixth of its diameter	Volume.
Circumference of a sphere × its diameter	Surface.
Square of the diameter of a sphere × 3.1416	Surface.
Square of the circumference of a sphere × 0.3183	Surface.
Cube of the diameter of a sphere × 0.5236	Volume.
Cube of the circumference of a sphere × 0.016887	Volume.
Radius of a sphere × 1.1547	Side of an inscribed cube.
Diameter of a sphere divided by ÷3	Side of an inscribed cube.
Area of its base × one-third of its altitude	Volume of a cone or pyramid whether round, square or triangular.
Area of one of its sides × 6	Surface of the cube.
Altitude of trapezoid × one-half the sum of its parallel sides	Area.

DECIMAL EQUIVALENTS OF FRACTIONS

8ths	32nds	64ths	64ths
$\frac{1}{8}$ = .125	$\frac{1}{32}$ = .03125	$\frac{1}{64}$ = 0.15625	$\frac{33}{64}$ = .515625
$\frac{1}{4}$ = .250	$\frac{3}{32}$ = .09375	$\frac{3}{64}$ = .046875	$\frac{35}{64}$ = .546875
$\frac{3}{8}$ = .375	$\frac{5}{32}$ = .15625	$\frac{5}{64}$ = .078125	$\frac{37}{64}$ = .57812
$\frac{1}{2}$ = .500	$\frac{7}{32}$ = .21875	$\frac{7}{64}$ = .109375	$\frac{39}{64}$ = .609375
$\frac{5}{8}$ = .625	$\frac{9}{32}$ = .28125	$\frac{9}{64}$ = .140625	$\frac{41}{64}$ = .640625
$\frac{3}{4}$ = .750	$\frac{11}{32}$ = .34375	$\frac{11}{64}$ = .171875	$\frac{43}{64}$ = .671875
$\frac{7}{8}$ = .875	$\frac{13}{32}$ = .40625	$\frac{13}{64}$ = .203128	$\frac{45}{64}$ = .703125
16ths	$\frac{15}{32}$ = .46875	$\frac{15}{64}$ = .234375	$\frac{47}{64}$ = .734375
$\frac{1}{16}$ = .0625	$\frac{17}{32}$ = .53125	$\frac{17}{64}$ = .265625	$\frac{49}{64}$ = .765625
$\frac{3}{16}$ = .1875	$\frac{19}{32}$ = .59375	$\frac{19}{64}$ = .296875	$\frac{51}{64}$ = .796875
$\frac{5}{16}$ = .3125	$\frac{21}{32}$ = .65625	$\frac{21}{64}$ = .328125	$\frac{53}{64}$ = .828125
$\frac{7}{16}$ = .4375	$\frac{23}{32}$ = .71875	$\frac{23}{64}$ = .359375	$\frac{55}{64}$ = .859375
$\frac{9}{16}$ = .5625	$\frac{25}{32}$ = .78125	$\frac{25}{64}$ = .390625	$\frac{57}{64}$ = .890625
$\frac{11}{16}$ = .6875	$\frac{27}{32}$ = .84375	$\frac{27}{64}$ = .421875	$\frac{59}{64}$ = .921875
$\frac{13}{16}$ = .8125	$\frac{29}{32}$ = .90625	$\frac{29}{64}$ = .453125	$\frac{61}{64}$ = .953125
$\frac{15}{16}$ = .9375	$\frac{31}{32}$ = .96875	$\frac{31}{64}$ = .484375	$\frac{63}{64}$ = .984375

CONVERSION TABLE FOR TEMP. – °F / °C

°F	°C	°F	°C	°F	°C
−459.4	−273	111.2	44	482	250
−418.0	−250	114.8	46	500	260
−328.0	−200	118.4	48	518	270
−238.0	−150	122.0	50	536	280
−193.0	−125	125.6	52	554	290
−148.0	−100	129.2	54	572	300
−130.0	−90	132.8	56	590	310
−112.0	−80	136.4	58	608	320
−94.0	−70	140.0	60	626	330
−76.0	−60	143.6	62	644	340
−58.0	−50	147.2	64	662	350
−40.0	−40	150.8	66	680	360
−36.4	−38	154.4	68	698	370
−32.8	−36	158.0	70	716	380
−29.2	−34	161.6	72	734	390
−25.6	−32	165.2	74	752	400
−22.0	−30	168.8	76	788	420
−18.4	−28	172.4	78	824	440
−14.8	−26	176.0	80	860	460
−11.2	−24	179.6	82	896	480
−7.6	−22	183.2	84	932	500
−4.0	−20	186.8	86	968	520
−0.4	−18	190.4	88	1004	540
3.2	−16	194.0	90	1040	560
6.8	−14	197.6	92	1076	580
10.4	−12	201.2	94	1112	600
14.0	−10	204.8	96	1202	650
17.6	−8	208.4	98	1292	700
21.2	−6	212.0	100	1382	750
24.8	−4	221.0	105	1472	800
28.4	−2	230.0	110	1562	850
32.0	0	239.0	115	1652	900
35.6	2	248.0	120	1742	950
39.2	4	257.0	125	1832	1000
42.8	6	266.0	130	2732	1500
46.4	8	275.0	135	3632	2000
50.0	10	284.0	140	4532	2500
53.6	12	293.0	145	5432	3000
57.2	14	302.0	150	6332	3500
60.8	16	311.0	155	7232	4000
64.4	18	320.0	160	4500	8132
68.0	20	329.0	165	9032	5000
71.6	22	338.0	170	9932	5500
75.2	24	347.0	175	10832	6000
78.8	26	356.0	180	11732	6500
82.4	28	365.0	185	12632	7000
86.0	30	374.0	190	13532	7500
89.6	32	383.0	195	14432	8000
93.2	34	392.0	200	15332	8500
96	36	410	210	16232	9000
100.4	38	428	220	17132	9500
104.0	40	446	230	18032	10000
107.6	42	464	240		

REFERENCE

COMMON ENGINEERING UNITS AND THEIR RELATIONSHIP

Quantity	SI Metric Units/Symbols	Customary Units	Relationship of Units
Acceleration	meters per second squared (m/s^2)	feet per second squared (ft/s^2)	$m/s^2 = ft/s^2 \times 3.281$
Area	square meter (m^2) square millimeter (mm^2)	square foot (ft^2) square inch (in^2)	$m^2 = ft^2 \times 10.764$ $mm^2 = in^2 \times 0.00155$
Density	kilograms per cubic meter (kg/m^3) grams per cubic centimeter (g/cm^3)	pounds per cubic foot (lb/ft^3) pounds per cubic inch (lb/in^3)	$kg/m^3 = lb/ft^2 \times 16.02$ $g/cm^3 = lb/in^2 \times 0.036$
Work	Joule (J)	foot pound force (ft lbf or ft lb)	$J = ft\ lbf \times 1.356$
Heat	Joule (J)	British thermal unit (Btu) Calorie (Cal)	$J = Btu \times 1.055$ $J = cal \times 4.187$
Energy	kilowatt (kW)	Horsepower (HP)	$kW = HP \times 0.7457$
Force	Newton (N) Newton (N)	Pound-force (lbf, lb · f, or lb) kilogram-force (kgf, kg · f, or kp)	$N = lbf \times 4.448$ $N = \dfrac{kgf}{9.807}$
Length	meter (m) millimeter (mm)	foot (ft) inch (in)	$m = ft \times 3.281$ $mm = \dfrac{in}{25.4}$
Mass	kilogram (kg) gram (g)	pound (lb) ounce (oz)	$kg = lb \times 2.2$ $g = \dfrac{oz}{25.35}$
Stress	Pascal = Newton per second (Pa = N/s)	pounds per square inch (lb/in^2 or psi)	$Pa = lb/in^2 \times 6,895$
Temperature	degree Celsius (°C)	degree Fahrenheit (°F)	$°c = \dfrac{°F - 32}{1.8}$
Torque	Newton meter (N · m)	foot-pound (ft lb) inch-pound (in lb)	$N \cdot m = ft\ lbf \times 1.356$ $N \cdot m = in\ lbf \times 0.113$
Volume	cubic meter (m3) cubic centimeter (cm3)	cubic foot (ft^3) cubic inch (in^3)	$m^3 = ft^3 \times 35.314$ $cm^3 = \dfrac{in^3}{16.387}$

METRIC INFORMATION		
BASE UNITS		
QUANTITY	**UNIT**	**SYMBOL**
Length	Meter	m
Mass	Kilogram	kg
Time	Second	s
Electric Current	Ampere	A
Thermodynamic Temperature	Kelvin	K
Amount of Substance	Mole	mo
Luminous Intensity	Candela	cd
SUPPLEMENTARY SI UNITS		
QUANTITY	**UNIT**	**SYMBOL**
Plane Angle	Radian	rad
Solid Angle	Steradian	sr

DERIVED METRIC UNITS WITH COMPOUND NAMES

PHYSICAL QUANTITY	**UNIT**	**SYMBOL**
Area	Square Meter	m^2
Volume	Cubic Meter	m^3
Density	Kilogram per Cubic Meter	kg/m^3
Velocity	Meter per Second	m/s
Angular Velocity	Radian per Second	rad/s
Acceleration	Meter per Second Squared	m/s^2
Angular Acceleration	Radian per Second Squared	rad/s^2
Volume Rate of Flow	Cubic Meter per Second	m^3/s
Moment of Inertia	Kilogram Meter Squared	$kg{\bullet}m^2$
Moment of Force	Newton Meter	N•m
Intensity of Heat Flow	Watt per Square Meter	W/m^2
Thermal Conductivity	Watt per Meter Kelvin	W/m•K
Luminance	Candela per Square Meter	cd/m^2

METRIC INFORMATION		
SI PREFIXES		
MULTIPLICATION FACTOR	**PREFIX**	**SYMBOL**
1 000 000 000 000 000 000 = 10^{18}	exa	E
1 000 000 000 000 000 = 10^{15}	peta	P
1 000 000 000 000 = 10^{12}	tera	T
1 000 000 000 = 10^9	giga	G
1 000 000 = 10^6	mega	M
1000 = 10^3	kilo	k
100 = 10^2	hecto	h
10 = 10^1	deka	da
0.1 = 10^{-1}	deci	d
0.01 = 10^{-2}	centi	c
0.001 = 10^{-3}	milli	m
0.000 001 = 10^{-6}	micro	m
0.000 000 001 = 10^{-9}	nano	n
0.000 000 000 001 = 10^{-12}	pico	p
0.000 000 000 000 001 = 10^{-15}	femto	f
0.000 000 000 000 000 001 = 10^{-18}	atto	a

REFERENCE

METRIC UNIT TO IMPERIAL UNIT CONVERSION FACTORS	
Metric Units	**Imperial Equivalents**
LENGTH	
1 Millimeter (mm)	= 0.0393701 Inch
1 Meter (m)	= 39.3701 Inches = 3.28084 Feet
1 Kilometer (km)	= 0.621371 Mile
LENGTH/TIME	
1 Meter per Second (m/s)	= 3.28084 Feet per Second
1 Kilometer per Hour (km/h)	= 0.621371 Mile per Hour
AREA	
1 Square Millimeter (mm²)	= 0.001550 Square Inch
1 Square Meter (m²)	= 19.7639 Square Feet
1 Hectare (ha)	= 2.47105 Acres
1 Square Kilometer (km²)	= 0.386102 Square Mile
VOLUME	
1 Cubic Millimeter (mm³)	= 0.0000610237 Cubic Inch
1 Cubic Meter (m³)	= 35.3147 Cubic Feet = 1.30795 Cubic Yards
1 Milliliter (ml.)	= 0.0351951 Fluid Ounce
1 Liter (L)	= 0.219969 Gallon
MASS	
1 Gram (g)	= 0.0352740 Ounce
1 Kilogram (kg)	= 2.20462 Pounds
1 Tonne (t) (51,000 kg)	= 1.10231 Tons (2,000 lb.)
FORCE	
1 Newton (N)	= 0.224809 Pound-Force
STRESS	
1 Megapascal (MPa)	= 145.038 Pounds-Force psi
LOADING	
1 Kilonewton per Sq. Meter	= 20.8854 Pounds Force psf
1 Kilonewton per Meter	= 68.5218 Pounds Force per Ft.
MISCELLANEOUS	
1 Joule (J)	= 0.00094781 B.t.u.
1 Joule (J)	= 1 Watt-second
1 Watt (W)	= 0.00134048 Electric hp

IMPERIAL UNIT TO METRIC UNIT CONVERSION FACTORS	
Imperial Units	**Metric Equivalents**
LENGTH	
1 Inch	= 25.4 mm = 0.0254 m
1 Foot	= 0.3048 m
1 Mile	= 1.60934 km
LENGTH/TIME	
1 Foot per Second	= 0.3048 m/s
1 Mile per Hour	= 1.60934 km/h
AREA	
1 Square Inch	= 645.16 mm²
1 Square Foot	= 0.0929030 m²
1 Acre	= 0.404686 ha
1 Square Mile	= 2.58999 km²
VOLUME	
1 Cubic Inch	= 16387.1 mm³
1 Cubic Foot	= 0.0283168 m³
1 Cubic Yard	= 0.764555 m³
1 Fluid Ounce	= 28.4131 mL
1 Gallon	= 4.54609 L
MASS	
1 Ounce	= 28.3495 g
1 Pound	= 0.453592 kg
1 Ton	= 0.907185 t
FORCE	
1 Pound	= 4.44822 N
STRESS	
1 psi	= 0.00689476 MPa
LOADING	
1 psf	= 0.0478803 kN/m²
1 plf	= 0.0145939 kN/m
MISCELLANEOUS	
1 B.t.u.	= 1055.06 J
1 Watt-second	= 1 J
1 Horsepower	= 746 W

ABBREVIATIONS USED ON ARCHITECTURAL DRAWINGS

A

Above Finished Counter	**AFC**
Above Finished Floor	**AFF**
Above Finished Grade	**AFG**
Acoustic	**AC**
Acoustic Plaster	**AC PL**
Acoustic Tile	**AC T**
Actual	**ACT**
Additional	**ADD**
Adhesive	**ADH**
Adjustable	**ADJ**
Aggregate	**AGGR**
Air Conditioning	**AIR COND**
Air Conditioning Unit	**ACU**
Alternating Current	**AC**
Aluminum	**AL or ALUM**
Amount	**AMT**
Ampere	**AMP or A**
Anchor Bolt	**AB**
Angle (in degrees)	∢
Angle (structural)	**L**
Approximate	**APPROX**
Architectural	**ARCH**
Area	**A**
Area Drain	**AD**
Asbestos	**ASB**
Asphalt	**ASPH**
Asphaltic Concrete	**ASPH CONC**
Assembly	**ASSEM**
At	**@**
Automatic	**AUTO**
Avenue	**AVE**
Average	**AVG**

B

Balcony	**BALC**
Basement	**BSMT**
Baseplate	**BP**
Bathroom	**B**
Bathtub with Shower	**BTS**
Batten	**BATT**
Beam	**BM**
Beam, Standard	**S BM**
Beam, Wide Flange	**W BM**
Bearing	**BRG**
Bearing Plate	**B PL**
Bedroom	**BR**
Bench Mark	**BM**
Between	**BET**
Beveled	**BEV**
Bidet	**BDT**
Block	**BLK**
Blocking	**BLKG**
Blower	**BLO**
Board	**BD**
Board Feet	**BD FT**
Both Sides	**BS**
Both Ways	**BW**
Bottom	**BOT**
Boulevard	**BLVD**
Bracket	**BRKT**
Brass	**BR**
Brick	**BRK**
British Thermal Unit	**BTU**
Broom Closet	**BC**
Building	**BLDG**
Building Line	**BL**
Built-in	**BLT-IN**
Built-up	**BU**
Buzzer	**BUZ**
By (used as 2 × 4)	×

C

Cabinet	**CAB**
Candela	**cd**
Candlepower	**CP**
Carpet	**CPT**
Cast Iron	**CI**
Cast in Place	**CIP**
Catch Basin	**CB**
Caulking	**CLKG**
Ceiling	**CLG**
Ceiling Diffuser	**CD**
Celsius	**C**
Cement	**CEM**
Cement Plaster	**CEM PLAS**
Center	**CTR**
Center to Center	**C to C**
Centerline	C_L **or CL**
Centimeter	**cm**
Ceramic	**CER**
Ceramic Tile	**CT**
Chalkboard	**CHKBD**
Chamber	**CHAM**
Channel (structural)	**C**
Check	**CHK**
Cinder Block	**CLN BL**
Circle	**CIR**
Circuit	**CKT**
Circuit Breaker	**CIR BKR**
Class	**CL**
Classroom	**CLRM**
Cleanout	**CO**
Clear	**CLR**
Closet	**CLO or CL**
Clothes Dryer	**CL D**
Cold Water	**CW**
Column	**COL**
Combination	**COMB**
Common	**COM**
Concrete	**CONC**
Concrete Block	**CONC B**
Concrete Masonry Unit (concrete block)	**CMU**
Construction	**CONST**
Continuous	**CONT**
Contractor	**CONTR**
Contractor Furnished	**CF**
Control Joint	**CJ**
Copper	**COP or CU**
Corridor	**CORR**
Counter	**CTR**
Countersink	**CSK**

ABBREVIATIONS USED ON ARCHITECTURAL DRAWINGS *(cont.)*

Courses	C	**E**		Firebrick	FBRK
Cover	COV	Each	EA	Fire-extinguisher	F EXT
Cross Section	X-SECT	Each Face	EF	Fire-extinguisher Cabinet	FEC
Cubic	CU	Each Way	EW	Fire Hose Cabinet	FHC
Cubic Feet	CU FT	East	E	Fire Hydrant	FH
Cubic Feet per Minute	CFM	Elbow	ELL	Fireproof	FP
Cubic Yard	CU YD	Electric(al)	ELECT	Fitting	FTG
D		Electric Panel Board	EPB	Fixture	FIX
Damper	DMPR	Elevation	EL or ELEV	Flammable	FLAM
Decibel	db	Elevator	ELEV	Flange	FLG
Deep, Depth	DP	Enclosure	ENCL	Flashing	FL
Degree	° or DEG	Engineer	ENGR	Flexible	FLEX
Department	DEPT	Entrance	ENT	Floor	FLR
Detail	DET	Equal	EQ	Floor Drain	FD
Diagonal	DIAG	Equipment	EQUIP	Floor Sink	FS
Diagram	DIAG	Estimate	EST	Flooring	FLG
Diameter	DIA	Excavate	EXC	Fluorescent	FLUOR
Diffuser	DIFF	Exhaust	EXH	Folding	FLDG
Dimension	DIM	Existing	EXIST'G	Foot	' or FT
Dining Room	DIN RM	Expansion Bolt	EB	Footing	FTG
Direct Current	DC	Expansion Joint	EXP JT	Forward	FWD
Dishwasher	DW	Exposed	EXPO	Foundation	FND
Disposal	DISPL	Extension	EXT	Four-way	4-W
Distance	DIST	Exterior	EXT	Frame	FR
Ditto	DO	Exterior Grade	EXT GR	Front	FR
Divided or Division	DIV	**F**		Full Size	FS
Door	DR	Fabricate	FAB	Furnace	FURN
Double	DBL	Face Brick	FB	Future	FUT
Double-hung	DH	Face of Studs	FCS	**G**	
Double-strength (glass)	DS	Fahrenheit	F	Gallon	GAL
Douglas Fir	DF	Feet	' or FT	Galvanized	GALV
Dowel	DWL	Feet per Minute	FPM	Galvanized Iron (galvanized steel)	GI
Down	DN	Fiberglass-reinforced Plastic	FRP	Gauge	GA
Downspout	DS	Figure	FIG	Glass	GL
Drain	D or DR	Finish(ed)	FIN	Glass block	GL BL
Drawing	DWG	Finished All Over	FAO	Glazed Structural Unit	GSU
Drinking Fountain	DF	Finished Floor	FIN FL		
Dryer	D	Finished Floor Elevation	FFE	Glue-laminated	GLUELAM
Drywall	DW	Finished Grade	FIN GR	Government	GOVT
Duplicate	DUP	Finished Opening	FO	Grade	GR

ABBREVIATIONS USED ON ARCHITECTURAL DRAWINGS

Grade Beam	**GB**	Install	**INST**	Longitude	**LNG**
Grating	**GRTG**	Insulate(d)(ion)	**INS**	Lumber	**LBR**
Gravel	**GVL**	Interior	**INT**	**M**	
Grille	**GR**	Interior Grade	**INT GR**	Manhole	**MH**
Ground	**GRND**	**J**		Manufacture(r)	**MFR**
Grout	**GT**	Jamb	**JMB**	Marble	**MRB**
Gypsum	**GYP**	Janitor's Sink	**JS**	Mark	**MK**
H		Janitor's Closet	**JC**	Masonry	**MAS**
Hall	**H**	Joint	**JT**	Masonry Opening	**MO**
Hardboard	**HBD**	Joist	**JST**	Material	**MAT**
Hardware	**HDW**	Joist and Plank	**J & P**	Maximum	**MAX**
Hardwood	**HDWD**	Junction	**JCT**	Mechanical	**MECH**
Head	**HD**	Junction Box	**J-BOX**	Medicine Cabinet	**MC**
Header	**HDR**	**K**		Medium	**MED**
Heater	**HTR**	Kelvin	**K**	Membrane	**MEMB**
Heating	**HTG**	Kiln Dried	**KD**	Metal	**MET**
Heating/Ventilating/	**HVAC**	Kilogram	**kg**	Metal Lath and	**MLP**
Air Conditioning		Kilovolt	**KV**	Plaster	
Heavy Duty	**HD**	Kilowatt	**KW**	Meter	**m**
Height	**HT**	Kitchen	**KIT**	Millimeter	**mm**
Hexagonal	**HEX**	Kitchen Cabinet	**KCAB**	Minimum	**MIN**
Highway	**HWY**	Kitchen Sink	**KSK**	Mirror	**MIRR**
Hollow Core	**HC**	Knockout	**KO**	Miscellaneous	**MISC**
Hollow Metal	**HM**	**L**		Modular	**MOD**
Horizontal	**HORIZ**	Laboratory	**LAB**	Molding	**MLDG**
Horsepower	**HP**	Laminate(d)	**LAM**	Mullion	**MULL**
Hose Bibb	**HB**	Landing	**LDG**	**N**	
Hospital	**HOSP**	Latitude	**LAT**	Noise Reduction	**NRC**
Hot Water	**HW**	Laundry	**LAU**	Coefficient	
Hot Water Heater	**HWH**	Lavatory	**LAV**	Nominal	**NOM**
Hour	**HR**	Left	**L**	North	**N**
House	**HSE**	Length	**LGTH**	Not Applicable	**NA**
Hundred	**C**	Level	**LEV**	Not in Contract	**NIC**
I		Library	**LIB**	Not to Scale	**NTS**
Illuminate	**ILLUM**	Light (pane of glass)	**LT**	Number	**NO. or #**
Incandescent	**INCAND**	Linear Feet	**LIN FT**	**O**	
Inch(es)	**" or IN.**	Linen Cioset	**L CL**	Oak	**O**
Inflammable	**INFL**	Linoleum	**LINO**	Office	**OFF**
Information	**INFO**	Live Load	**LL**	On Center	**OC**
Inside Diameter	**ID**	Living Room	**LR**	One-way	**1-W**
Inside Face	**IF**	Location	**LOC**	Open Web	**OW**
Inspect(ion)	**INSP**	Long	**LG**	Opening	**OPG**

REFERENCE

Opposite	**OPP**	Pounds per Square Inch	**PSI**	Roof Drain	**RD**
Opposite Hand	**OPH**			Roofing	**RFG**
Ounce	**OZ**	Precast	**PRCST**	Room	**RM**
Outside Diameter	**OD**	Prefabricated	**PREFAB**	Rough	**RGH**
Outside Face of Concrete	**OFC**	Preliminary	**PRELIM**	Rough Opening	**RO**
		Premolded	**PRMLD**	Round	**RD or Ø**
Outside Face of Studs	**OFS**	Property	**PROP**	Rubber Base	**RB**
		Public Address System	**PA**	Rubber Tile	**RBT**
Overhead	**OH**			**S**	
P		Pull Chain	**PC**	Schedule	**SCH**
Painted	**PTD**	Pushbutton	**PB**	Screw	**SCR**
Pair	**PR**	**Q**		Second	**s or SEC**
Panel	**PNL**	Quantity	**QTY**	Section	**SECT**
Parallel	**PAR or ‖**	Quarry Tile	**QT**	Select	**SEL**
Partition	**PTN**	Quart	**QT**	Select Structural	**SS**
Passage	**PASS**	**R**		Self-closing	**SC**
Pavement	**PVMT**	Radiator	**RAD**	Service	**SERV**
Penny (nail size)	**d**	Radius	**RAD**	Sewer	**SEW**
Per	**/**	Random Length and Width	**RL&W**	Sheathing	**SHTHG**
Percent	**%**	Range	**R**	Sheet	**SHT**
Perforate	**PERF**	Receptacle	**RECP**	Sheet Metal	**SM**
Perimeter	**PERIM**	Recessed	**REC**	Shower	**SH**
Perpendicular	**PERP or ⊥**	Redwood	**RDWD**	Siding	**SDG**
Pierce	**PC**	Reference	**REF**	Sill Cock	**SC**
Plan	**PLN**	Refrigeration	**REF**	Similar	**SIM**
Plaster	**PLS**	Refrigerator	**REFRIG**	Single-hung	**SH**
Plasterboard	**PL BD**	Register	**REG**	Single-strength (glass)	**SS**
Plastic	**PLAS**	Reinforced, Reinforcing	**REINF**		
Plastic Tile	**PLAS T**			Sink	**SK**
Plate	**PL or P$_L$**	Reinforcing Bar	**REBAR**	Slop Sink	**SS**
Plate Glass	**PL GL**	Required	**REQ**	Socket	**SOC**
Platform	**PLAT**	Resilient	**RES**	Soil Pipe	**SP**
Plumbing	**PLMB**	Resistance	**RES**	Solid Block	**SLD BLK**
Plywood	**PLY**	Return	**RET**	Solid Core	**SC**
Polished	**POL**	Revision	**REV**	South	**S**
Polyethelyne	**POLY or PE**	Revolutions per Minute	**RPM**	Specifications	**SPEC**
				Square	**□ or SQ**
Polystyrene	**PS**	Right	**R**	Square Feet	**SF or ⊡**
Polyvinyl Chloride	**PVC**	Right hand	**RH**	Square Inches	**SQ IN or ⊡**
Position	**POS**	Riser	**R**		
Pound	**LB or #**	Road	**RD**	Stainless Steel	**SST**
Pounds per Square Foot	**PFS**	Roof	**RF**	Stairs	**ST**

ABBREVIATIONS USED ON ARCHITECTURAL DRAWINGS

Stand Pipe	ST P	Terrazzo	TZ	Vinyl Tile	VT
Standard	STD	Thermostat	THERMO	Vinyl Wall Covering	VWC
Station Point	SP	Thickness	THK	Vitreous Clay Tile	VCT
Steel	STL	Thousand	M	Volt	V
Stirrup	STIR	Thousand Board Feet	MBM	Volume	VOL
Stock	STK			**W**	
Storage	STO	Three-way	3-W	Wainscot	WSCT
Storm Drain	SD	Threshold	THR	Wall Cabinet	VCAB
Street	ST	Toilet	TOL	Wall Vent	WV
Structural	STR	Tongue and Groove	T & G	Waste Stack	WS
Structural Clay Tile	SCT	Top of wall	TW	Water	W
Substitute	SUB	Tread	TR	Water Closet (toilet)	WC
Supply	SUP	Two-way	2-W	Water Heater	WH
Surface	SUR	Typical	TYP	Waterproof	WP
Surface Four Sides	S4S	**U**		Watt	W
Surface Two Edges	S2E	Undercut Door	UCD	Weatherproof	WP
Suspended Ceiling	SUSP CLG	Underwriters' Laboratory, Inc.	U.L.	Weephole	WH
Switch	S or SW			Weight	WT
Symbol	SYM	Unfinished	UNFIN	Welded Wire Fabric	WWF
Symmetrical	SYM	Urinal	UR	West	W
Synthetic	SYN	Utility	UTIL	Wet Bulb	WB
System	SYS	**V**		White Pine	WP
T		V-joint	VJ	Wide Flange (structural)	W
Tack Board	TK BD	Vanishing Point	VP	Window	WDW
Tangent	TAN	Vanity	VAN	With	w/
Tar and Gravel	T & G	Vapor Barrier	VB	Without	WO
Technical	TECH	Vent Through Roof	VTR	Wood	WD
Tee	T	Vent Stack	VS	Working Point	WPT
Telephone	TEL	Ventilation	VENT	Wrought Iron	WI
Television	TV	Ventilator	V	**Y**	
Temperature	TEMP	Vertical	VERT	Yard	YD
Temporary	TEMP	Vertical Grain	VG	Yellow Pine	YP
Terra-cotta	TC	Vestibule	VEST	**Z**	
		Vinyl	VIN	Zinc	ZN
		Vinyl Base	VB		

GRAPHIC SYMBOLS USED ON ARCHITECTURAL DRAWINGS

Following are some of the most frequently used graphic symbols. Variations of these can be found on drawings made by those working in the various areas presented on such drawings.

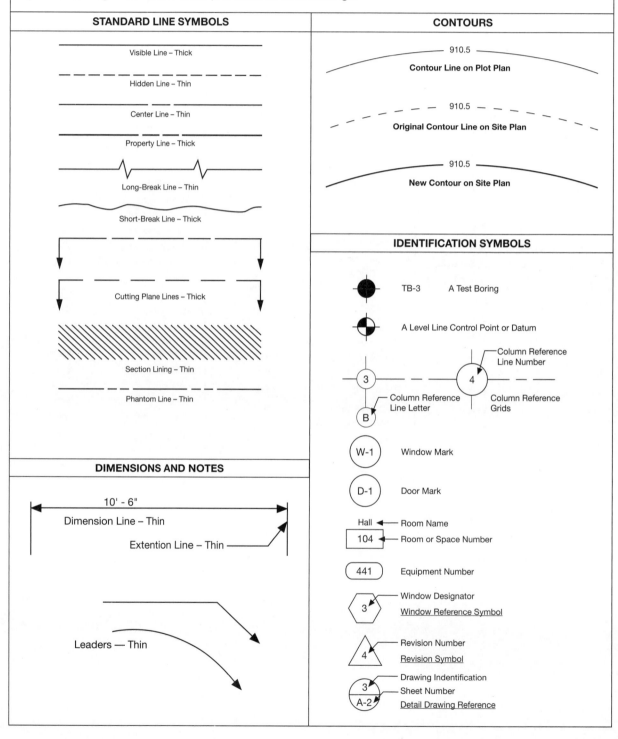

STANDARD LINE SYMBOLS

Visible Line – Thick

Hidden Line – Thin

Center Line – Thin

Property Line – Thick

Long-Break Line – Thin

Short-Break Line – Thick

Cutting Plane Lines – Thick

Section Lining – Thin

Phantom Line – Thin

DIMENSIONS AND NOTES

10' - 6"

Dimension Line – Thin

Extention Line – Thin

Leaders — Thin

CONTOURS

910.5

Contour Line on Plot Plan

910.5

Original Contour Line on Site Plan

910.5

New Contour on Site Plan

IDENTIFICATION SYMBOLS

TB-3   A Test Boring

A Level Line Control Point or Datum

Column Reference Line Number

3   4

Column Reference Line Letter

B   Column Reference Grids

W-1   Window Mark

D-1   Door Mark

Hall ◀ Room Name

104 ◀ Room or Space Number

441   Equipment Number

Window Designator

3   Window Reference Symbol

Revision Number

4   Revision Symbol

Drawing Indentification

3   Sheet Number

A-2   Detail Drawing Reference

GRAPHIC SYMBOLS USED ON ARCHITECTURAL DRAWINGS *(cont.)*

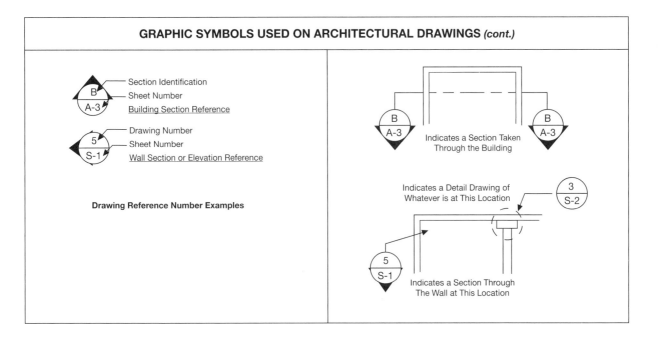

Section Identification
Sheet Number
Building Section Reference

Drawing Number
Sheet Number
Wall Section or Elevation Reference

Drawing Reference Number Examples

Indicates a Section Taken Through the Building

Indicates a Detail Drawing of Whatever is at This Location

Indicates a Section Through The Wall at This Location

SYMBOLS FOR MATERIALS IN SECTIONS

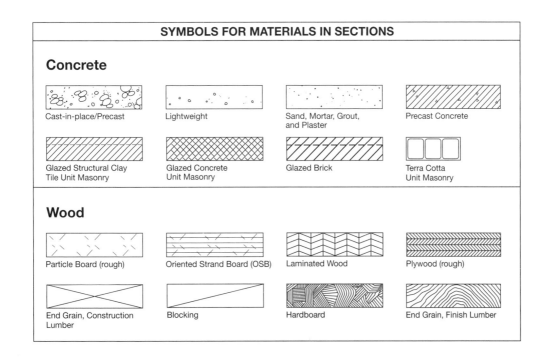

Concrete

Cast-in-place/Precast

Lightweight

Sand, Mortar, Grout, and Plaster

Precast Concrete

Glazed Structural Clay Tile Unit Masonry

Glazed Concrete Unit Masonry

Glazed Brick

Terra Cotta Unit Masonry

Wood

Particle Board (rough)

Oriented Strand Board (OSB)

Laminated Wood

Plywood (rough)

End Grain, Construction Lumber

Blocking

Hardboard

End Grain, Finish Lumber

REFERENCE

SYMBOLS FOR MATERIALS IN SECTIONS *(cont.)*

Masonry

Gypsum Unit Masonry

Cast Stone

Common/Face

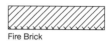
Fire Brick

Concrete Block

Structural Facing File

Tile Structural Clay

Clay Tile

Rough Cut Stone

Metal

Cast Iron

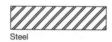
Steel

Aluminum

Brass/Bronze

Glass

Glass

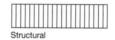
Structural

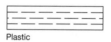
Plastic

Finish Materials

Gypsum Wallboard

Lath and Plaster

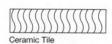
Ceramic Tile

Resilient Tile

Carpet and Pad

Terrazzo

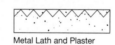
Metal Lath and Plaster

Marble

Slate, Bluestone, Flagging,
Soapstone

Site Work

Fine Porous Fill

Coarse Porous Fill (gravel)

Earth

Earth (alternate)

Rock

Rubble

SYMBOLS FOR MATERIALS IN SECTIONS *(cont.)*

Thermal Protection

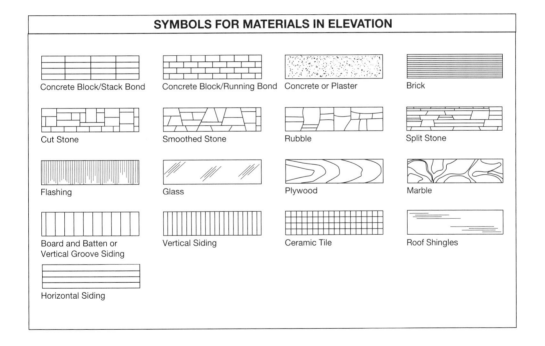

Insulation, Spray and Foam   Fibrous Fire Safing   Insulation, Rigid   Insulation Batts or Loose

Foam Insulation   Exterior Insulation and Finish System (EIFS)

SYMBOLS FOR MATERIALS IN ELEVATION

Concrete Block/Stack Bond   Concrete Block/Running Bond   Concrete or Plaster   Brick

Cut Stone   Smoothed Stone   Rubble   Split Stone

Flashing   Glass   Plywood   Marble

Board and Batten or Vertical Groove Siding   Vertical Siding   Ceramic Tile   Roof Shingles

Horizontal Siding

SYMBOLS FOR WALLS IN SECTION

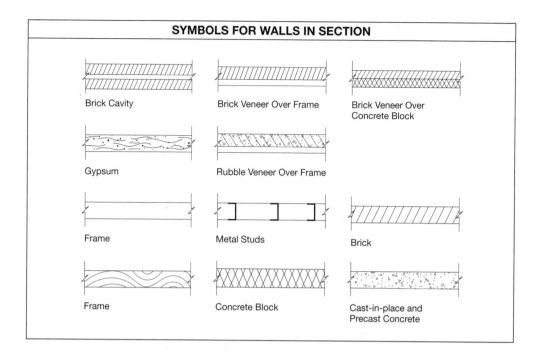

Brick Cavity

Brick Veneer Over Frame

Brick Veneer Over Concrete Block

Gypsum

Rubble Veneer Over Frame

Frame

Metal Studs

Brick

Frame

Concrete Block

Cast-in-place and Precast Concrete

LANDSCAPE SYMBOLS

PAVING

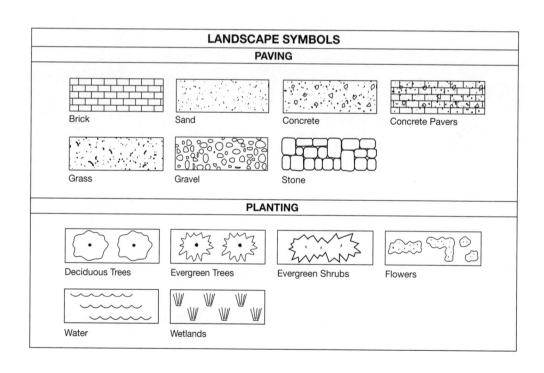

Brick

Sand

Concrete

Concrete Pavers

Grass

Gravel

Stone

PLANTING

Deciduous Trees

Evergreen Trees

Evergreen Shrubs

Flowers

Water

Wetlands

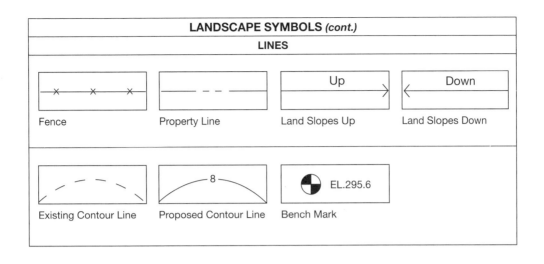

LANDSCAPE SYMBOLS *(cont.)*

LINES

Fence	Property Line	Land Slopes Up	Land Slopes Down
Existing Contour Line	Proposed Contour Line	Bench Mark	

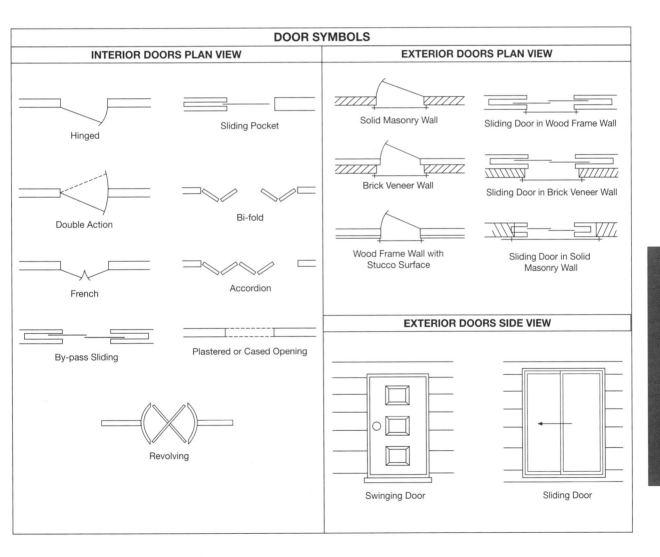

DOOR SYMBOLS

INTERIOR DOORS PLAN VIEW

Hinged

Sliding Pocket

Double Action

Bi-fold

French

Accordion

By-pass Sliding

Plastered or Cased Opening

Revolving

EXTERIOR DOORS PLAN VIEW

Solid Masonry Wall

Sliding Door in Wood Frame Wall

Brick Veneer Wall

Sliding Door in Brick Veneer Wall

Wood Frame Wall with Stucco Surface

Sliding Door in Solid Masonry Wall

EXTERIOR DOORS SIDE VIEW

Swinging Door

Sliding Door

REFERENCE

DOOR SYMBOLS (cont.)

Door Symbols in a Frame Wall

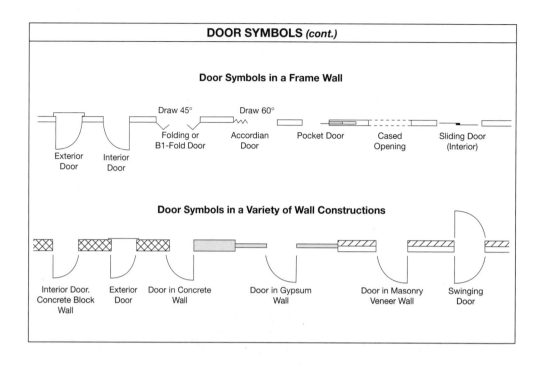

Door Symbols in a Variety of Wall Constructions

WINDOW SYMBOLS

Window Symbols in Masonry Wall

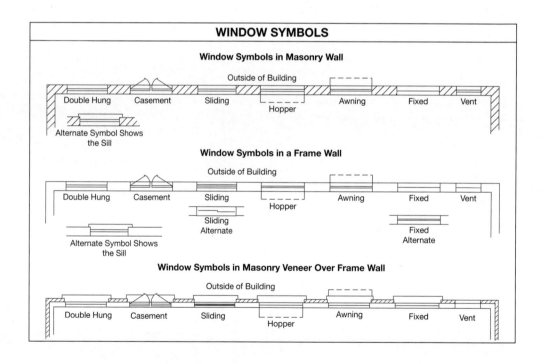

Window Symbols in a Frame Wall

Window Symbols in Masonry Veneer Over Frame Wall

WINDOW SYMBOLS *(cont.)*

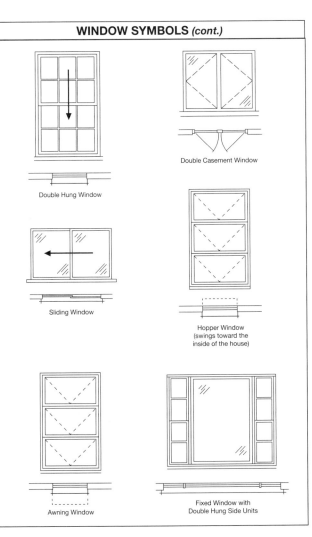

Double Hung Window

Sliding Window

Awning Window

Double Casement Window

Hopper Window
(swings toward the
inside of the house)

Fixed Window with
Double Hung Side Units

TYPICAL OPENINGS IN VARIOUS WALL CONTRUCTIONS

WOOD OR STEEL STUD EXTERIOR WALL

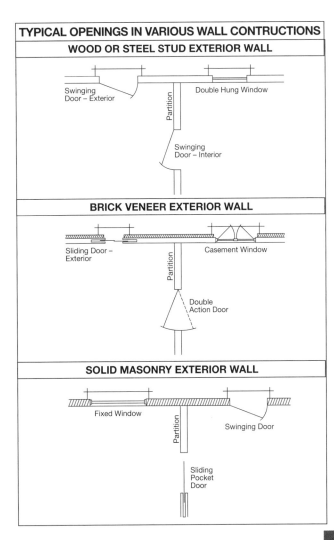

Swinging
Door – Exterior

Double Hung Window

Partition

Swinging
Door – Interior

BRICK VENEER EXTERIOR WALL

Sliding Door –
Exterior

Casement Window

Partition

Double
Action Door

SOLID MASONRY EXTERIOR WALL

Fixed Window

Partition

Swinging Door

Sliding
Pocket
Door

RECEPTACLE OUTLET SYMBOLS

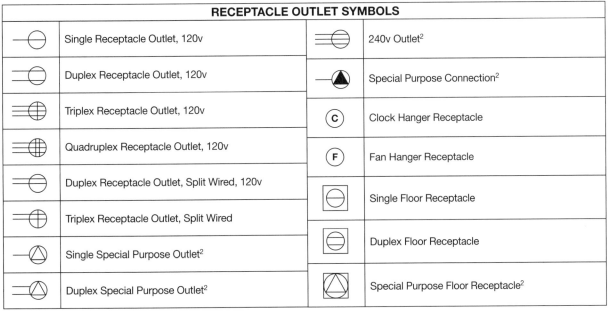

Symbol	Description	Symbol	Description
	Single Receptacle Outlet, 120v		240v Outlet[2]
	Duplex Receptacle Outlet, 120v		Special Purpose Connection[2]
	Triplex Receptacle Outlet, 120v	C	Clock Hanger Receptacle
	Quadruplex Receptacle Outlet, 120v	F	Fan Hanger Receptacle
	Duplex Receptacle Outlet, Split Wired, 120v		Single Floor Receptacle
	Triplex Receptacle Outlet, Split Wired		Duplex Floor Receptacle
	Single Special Purpose Outlet[2]		Special Purpose Floor Receptacle[2]
	Duplex Special Purpose Outlet[2]		

LIGHTING OUTLET SYMBOLS

Symbol	Description	Symbol	Description
◯	Incandescent fixture, surface or pendant	B	Emergency battery pack and charger, has sealed beams
R	Recessed incandescent fixture		Single fluorescent fixture, surface mounted
○○○○	Incandescent lighting track		Emergency service fluorescent fixture
D	Drop cord	R	Single recessed fluorescent fixture
Exit symbol	Exit light and outlet box, directional arrows to exit, shaded area is the front		Continuous row fluorescent fixture, surface mounted
J	Junction box	R	Recessed continuous row fluorescent fixture
L PS	Lamp holder with pull switch		Single bare lamp fluorescent fixture
L	Outlet controlled by low voltage switching when relay is in outlet box		Continuous bare lamp fluorescent fixture

1. Outlets requiring special identification may be indicated by lettering abbreviations beside the standard symbol, as WP for weatherproof or EP for explosion proof.

2. Use numeral or letter beside symbol keyed to a legend of symbols to indicate the type of receptacle or its use.

SWITCH SYMBOLS

S	Single Pole Switch
S_2	Double Pole Switch
S_3	Three-way Switch
S_4	Four-way Switch
S_P	Switch with Pilot Lamp
S_K	Key-operated Switch
S_L	Switch – Low Voltage System
S_{LM}	Master Switch – Low Voltage System
S_T	Time Switch
S_D	Door Switch
S_{DM}	Dimmer Switch
S_D	Automatic Door Switch
(S)	Ceiling Pull Switch
◯ S	Switch with Single Receptacle
◯ S	Switch with Double Receptacle
S_{CB}	Circuit Breaker Switch

SIGNALING AND COMMUNICATIONS SYMBOLS

Symbol	Description	Symbol	Description
•	Pushbutton		Buzzer
Bell symbol	Bell	D	Electric Door Opener
CH	Chime		Bell and Buzzer
	Interconnection Box	BT	Bell-Ringing Transformers
R	Radio Outlet	TV	Television Outlet
◀	Data Communications	◁	Telephone
Floor Telephone symbol	Floor Telephone	(S)	Speaker
F	Fire Alarm Pull Control	F ◁	Fire Alarm Strobe/Horn
F ◯	Fire Alarm Bell		

WIRE SYMBOLS

⌒	Wiring Concealed in Wall and Ceiling	- - -	Surface-mounted Wiring
- - -	Wiring Concealed in the Floor	———	Rigid Conduit
∿	Flexible Conduit	◀—	Home Run to Panel
—|—	Wiring, Neutral	—|—	Wiring, Hot
	Wiring, Ground	—•|—	

1-30

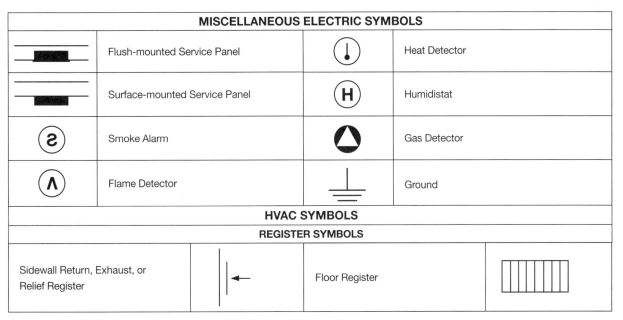

MISCELLANEOUS ELECTRIC SYMBOLS

	Flush-mounted Service Panel		Heat Detector
	Surface-mounted Service Panel	(H)	Humidistat
(S)	Smoke Alarm		Gas Detector
(Λ)	Flame Detector		Ground

HVAC SYMBOLS
REGISTER SYMBOLS

Sidewall Return, Exhaust, or Relief Register		Floor Register	

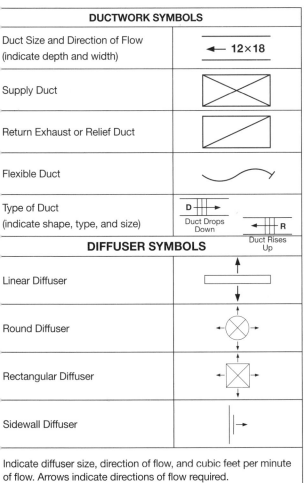

DUCTWORK SYMBOLS

Duct Size and Direction of Flow (indicate depth and width)	← 12×18
Supply Duct	
Return Exhaust or Relief Duct	
Flexible Duct	
Type of Duct (indicate shape, type, and size)	D — Duct Drops Down · R — Duct Rises Up

DIFFUSER SYMBOLS

Linear Diffuser	
Round Diffuser	
Rectangular Diffuser	
Sidewall Diffuser	

Indicate diffuser size, direction of flow, and cubic feet per minute of flow. Arrows indicate directions of flow required.

OTHER HVAC SYMBOLS

Thermostat	(T)	Damper	
Humidistat	H	Furnace	FURN

HOT WATER SYSTEM SYMBOLS

Exposed Radiator	
Recessed Radiator	
Flush Enclosed Radiator	
Centrifugal Unit Heater	
Air Eliminator Value	
Strainer	
Thermometer	
Pressure Gauge and Cock	
Circulating Pump	(C) →
Relief Valve	
Thermostat, Electric	(T)
Thermostat, Pneumatic	T

REFERENCE

AIR CONDITIONING PIPING SYMBOLS	
Refrigerant Liquid	—— RL ——
Refrigerant Discharge	—— RD ——
Refrigerant Suction	—— RS ——
Condenser Water Supply	—CWS—
Condenser Water Return	—CWR—
Chilled Water Supply	—CHWS—
Chilled Water Return	—CHWR—
Make Up Water	—— MU ——
Drain	—— D ——
Brine Supply	—— B ——
Brine Return	—— BR ——

HEATING PIPING SYMBOLS	
Hot Water Heating Supply	—— HW ——
Hot Water Heating Return	—— HWR—
High Pressure Steam	—— HPS—
Medium Pressure Steam	—— MPS—
Low Pressure Steam	—— LPS—
High Pressure Return	—— HPR—
Medium Pressure Return	—— MPR—
Low Pressure Return	—— LPR—
Boiler Blow Off	—— BD ——
Condensate or Vacuum Pump Discharge	—— VPD ——
Make-up Water	—— MU ——
Fuel Oil Suction	—— FOS—
Fuel Oil Return	—— FOR—
Fuel Oil Vent	—— FOV—
Gas – Low Pressure	—— G ——
Gas – Medium Pressure	—— MG ——
Gas – High Pressure	—— HG ——
Arrow Indicates Direction of Flow	——▶——

PLUMBING PIPING SYMBOLS	
WASTE AND VENTS	
Soil, Waste, or Leader (above grade)	————
Soil, Waste, or Leader (below grade)	— — — –
Vent	– – – – – –
Combination Waste and Vent	—SV—
Acid Waste	—AW—
Acid Vent	– – –AV– – –
Indirect Drain	—— IW ——
Storm Drain	—— S ——
FIRE PIPING	
Fire Line	— F —— F —
Wet Standpipe	—WSP—
Dry Standpipe	—DSP—
Combination Standpipe	—CSP—
Main Supplies Sprinkler	—— S ——
Branch and Head Sprinkler	—o —— o—
WATER	
Cold Water	– – —— – –
Soft Cold Water	—SW—
Industrialized Cold Water	—ICW—
Chilled Drinking Water Supply	—DWS—
Chilled Drinking Water Return	—DWR—
Hot Water	– – – – – –
Hot Water Return	– – – – – – –
Sanitizing Hot Water Supply (180° F)	⁄ – – ⁄ – – ⁄
Sanitizing Hot Water Return (180° F)	⁄– – – ⁄ – – –
Industrialized Hot Water Supply	—— IHW ——
Industrialized Hot Water Return	—— IHR ——
Tempered Water Supply	—TWG—
Tempered Water Return	—TWR—

OTHER PIPING	
Compressed Air	—— A ——
Vacuum	—— V ——
Vacuum Cleaning	—— VC ——
Oxygen	—— O ——
Liquid Oxygen	—LOX—
Nitrogen	—— N ——
Liquid Nitrogen	—— LN ——
Nitrous Oxide	—— NO ——
Hydrogen	—— H ——
Helium	—— HE ——
Argon	—— AR ——
Liquid Petroleum Gas	—LPG—
Industrial Waste	—— INW ——

PLUMBING SYMBOLS	
VALVES	
Check Valve, Straightway	
Gate Valve	
Globe Valve	
Butterfly	
Solenoid	
Lock Shield	
2-way Automatic Control	
3-way Automatic Control	
Gas Cock	
Pressure Reducing Valve	

PLUMBING SYMBOLS (cont.)	
FITTINGS	
90° Elbow	
90° Elbow, Turned Down	
90° Elbow, Turned Up	
45° Elbow	
Straight Cross	
Straight Tee	
Straight Tee, Outlet Up	
Straight Tee, Outlet Down	
Union	
Concentric Reducer	
Eccentric Reducer	
Cap	
P-Trap	
House Trap	
Shock Absorber	

OTHER SYMBOLS	
Expansion Joint	
Expansion Loop	
Flexible Connection	
Flow Direction	
Pipe Pitch, Rise (R)/Drop (D)	
Thermostat	
Cleanout on End of Pipe	
Cleanout on Wall	
Floor, Roof, or Shower Drain	
Vent Through Roof	
1 1/2" Waste Down (up)	
1/2" Hot Water Down (up)	
1/2" Cold Water Down (up)	
Meter	
Air Vent	
Vent Pipe	

REFERENCE

SYMBOLS INDICATING TYPE OF PIPE MATERIAL

Cast Iron	——CI——
Clay Tile	——CT——
Ductile Iron	——DI——
Reinforced Concrete	——RCP——
Polyvinyl Chloride	——PVC——
Acrilylonitrile Butadiene Styrene	——ABS——
Styrene Rubber Plastic	——SRP——

SYMBOLS INDICATING MATERIALS CARRIED BY PIPE

Material	Symbol
Acid Waste	——AW——
Condensate	——C——
Compressed Air	——CA——
Carbon Dioxide	——CO2——
Cold Water	——CW——
Dry Standpipe	——DSP——
Fuel Oil	——FO——
Gasoline	——GAS——
Gas – Low Pressure	——G——
Gas – Medium Pressure	——MG——
Gas – High Pressure	——HG——
Hot Water	——HW——
Liquefied Propane Gas	——LPG——
Nitrogen	——N——
Natural Gas	——NG——
Nontoxic Industrial Waste	——NTW——
Oxygen	——O——
Refrigerant	——RFGT——
Steam	——STM——
Water	——W——
Wet Standpipe	——WSP——

SYMBOLS INDICATING FLANGED, SCREWED, WELDED, OR SOLDERED FITTINGS

Fitting	Flanged	Screwed	Welded (X or •)	Soldered
90° Elbow				
90° Elbow, Turned Down				
90° Elbow, Turned Up				
45° Elbow				
Straight Cross				
Straight Tee				
Straight Tee, Outlet Up				
Straight Tee, Outlet Down				
Union				
Check Valve, Straightway				
Gate Valve				
Globe Valve				

FIRE PROTECTION SYSTEM SYMBOLS

Fire Line Water Supply	——F——	Combination Standpipe	——CSP——
Upright Sprinkler Head	——○——	Fire Hydrants	
Pendent Sprinkler Head	——●——	Wall Mounted Fire Hydrant with Two Heads	
Wet Standpipe	——WSP——	Recessed Fire Hose Cabinet	FHC
Dry Standpipe	——DSP——	Surface Mounted Fire Hose Cabinet	FHC

WATERPROOFING MEMBRANES AND COATINGS

Areas where waterproofing is required include foundations, roofs, exterior wood and masonry walls, land exposed structural components, including steel. Methods for waterproofing include:

1. Applying a built-up bituminous membrane of felt and hot or cold tar pitch.

2. Applying a heavy coating, such as Portland cement plaster or a trowelable asphalt.

3. Bonding an elastomeric membrane to the wall.

4. Applying a thin film or coating to the exterior of the wall, such as liquid silicone or coal tar pitch.

5. Adding waterproofing admixtures to the concrete as it is mixed.

6. Applying a dry coating that will emulsify in place, such as bentonite clay.

WATERPROOFING MATERIALS IN GENERAL USE

Sheet Membranes	Composite Membranes
Butyl Ethyene Propylene Neoprene Polyethylene Polyvinyl Chloride	Elastomeric, Backed Polyethylene and 　Rubberized Bitumen Polyvinyl Chloride Backed Saturated Felts and Bitumen 　Coated
Liquid Membranes	**Applied Coating**
Butyl Urethane Polychlorene (neoprene) Polyurethane, Coal Tar	Acrylic, Silicone Asphalt Emulsions, 　Cut Backs Cementitious with 　Admixtures Epoxy, Bitumen Urethane, Bitumen Bitumen, Rubberized

BUILT-UP MEMBRANES	
Hot Applied	**Cold Applied**
Asphalt, Type I, II, III Coal Tar Pitch, Type B Felts, Saturated and 　Coated	Bitumen Emulsion Bitumen, Fiberated 　Cement Felts, Coated Bentonite Clay Fabric, Saturated Glass Fiber Mesh, 　Saturated Cementitious Membrane

REFERENCE

WATERPROOF TIPS

There are a number of ways to waterproof a foundation wall. The manufacturer of a system usually requires the contractor to employ a certified applicator if the manufacturer's guarantee is to be valid. Common systems include liquid membranes, sheet membranes, cementitious coating, built-up systems, and bentonite.

Liquid membranes are applied with a roller, trowel, or spray. The liquid solidifies into a rubbery coating. Different materials are available, such as polymer-modified asphalt and various polyurethane liquid membranes.

Sheet membranes tend to be self-adhering rubberized asphalt sheets, typically an assembly of multiple layers of bitumen and reinforcing materials. Some companies manufacture PVC and rubber butyl sheet membranes.

Cementitious products are available from building supply outlets. They are mixed on the site and are applied with a brush. Some have an acrylic additive available that improves bonding and makes the cementitious coating more durable. One disadvantage is that these coatings will not stretch if the foundation cracks, thus opening the possibility for leakage through cracks.

Built-up systems may be like the widely used hot tar and felt membrane. Alternate layers of hot tar and felt are bonded to the foundation. Usually at least three layers of felt are specified.

Bentonite is a clay material that expands when wet. It is available in sheets that are adhered to the foundation. As groundwater penetrates the clay, it swells many times its original volume, providing a permanent seal against water penetration.

Surface Preparation

Regardless of the type of system used, it is important to prepare the surface before application. This includes (1) drying the wall and footings, (2) removing the concrete form ties, making certain they break out inside the foundation so they do not penetrate the waterproof membrane, (3) cleaning the wall so it is free of all dirt or other loose material, and (4) sweeping the wall free of dust and mud film residue. A residue left when wet mud is wiped off and left to dry on the foundation can inhibit bonding. Finally, any openings around pipes or other items that penetrate the wall must be grouted.

Safety

Waterproofing presents some hazards that must be controlled. First is a possible cave-in of the soil, burying the workers. Normal shoring procedures should be observed. Many of the materials used are flammable and solvent-based, presenting a fire hazard. Workers should not smoke or use any tools that might cause ignition. Solvent fumes can be very harmful, and workers must wear respirators. Fumes are usually heavier than air and settle around the foundation in the excavated area. The solvents, asphalt, and other materials used may cause skin problems, so protective clothing, including gloves, is required with many products. As always, wear eye protection. When in doubt, consult the manufacturer of the product.

WEIGHTS OF CONSTRUCTION MATERIALS

BRICK AND MASONRY

Type	lb./sq. ft.
4" Brick Wall	40
4" Concrete Brick, Stone or Gravel	46
4" Concrete Brick, Lightweight	33
4" Concrete Block, Stone or Gravel	34
4" Concrete Block, Lightweight	22
6" Concrete, Stone or Gravel	50
6" Concrete Block, Lightweight	31
8" Concrete Block, Stone, or Gravel	55
8" Concrete Block, Lightweight	35
12" Concrete Block, Stone or Gravel	85
12" Concrete Block, Lightweight	55

CONCRETE

Type	lb./sq. ft.
Plain, Slag	132
Plain, Stone	144
Reinforced, Slag	138
Reinforced, Stone	150

GLASS

Type	lb./sq. ft.
Double Strength, 1/8"	26 ounces
Double Pane Insulating with 5/8" Air space	3.25
Glass Block, 4" Standard with Mortar	20
Glass Block, 3" Solid with Mortar	40
Glass Block, 3" Lightweight with Mortar	16
Wire Glass, 1/4"	18

SOFT WOODS

Type	lb./cu. ft. at 12% Moisture Content
Balsam Fir	23.8
Cedar, White	21.0
Cypress	33.0
Cedar, Western	24.5
Fir, Larch, Douglas	34.2
Hemlock	29.4
Pine, Yellow Southern	36.4
Pine, Northern	32.2
Pine, Ponderosa	29.4
Pine, White	25.9
Redwood, California	26.2
Spruce, Engleman	28.7

HARDWOODS

Type	lb./cu. ft. at 12% Moisture Content
Ash, White	40.5
Aspen	25.9
Birch	44.0
Cottonwood	28.0
Poplar, Yellow	30.1

REFERENCE

SPANISH SECTION CONTENTS

Pronouncing Spanish 2-1
   Vowels 2-1
   Dipthongs (double vowels) 2-1
   Consonants 2-1
   Plurarls 2-1

General Communication / Comunicación General 2-2
   Introduction / Introducción 2-2
   Getting Acquainted / Familiarizandose o Conociendose 2-2
   The Basics / Lo Básico 2-2

Hiring / Contratando 2-3
   Administration / Administración 2-3
   Verification / Verificación 2-3
   Prior to Hiring / Antes de Contratar 2-4
   Reporting to Work / Reportando a Trabajar 2-4
   Conduct on Job Site / Conducta en el lugar de trabajo 2-4
   Compliments / Cumplidos 2-4
   Code Administration / Codigo de Administración 2-4
   Construction Personnel / Personal de Construcción 2-4
   Administrative Documents / Documentos Administrativos 2-5

Site Work / Lugar de Trabajo 2-5

Tools & Equipment / Herramientas y Equipo 2-6
   Hand Tools / Herramientas Manuales 2-7
   Power Tools / Herramientas Eléctricas 2-7
   Landscape & Cleaning / Jardinería y Limpieza 2-8
   Motorized / Motorizado 2-8

Areas of a House or Building / Areas de la Casa o del Edificio 2-8

Safety / Seguridad 2-8
   Distress Situations / Situaciones de Peligro 2-8
   Injury Terms / Terminos para Lesiones 2-9
   Hazards / Riezgos 2-10
   Signs / Señales 2-10

Drywall / Tabla Roca o Tabla de Yeso 2-10

Paint / Pintura 2-11

Roof Covering / Revestimiento de Techo 2-12

Plumbing / Plomería 2-13
   General Terms / Términos Generales 2-13
   Toilet / Inodoro 2-14

SPANISH

Faucets / Grifos o Llaves 2-14
Tub/Shower Controls / Tina/Controles de la Ducha/Regadera 2-14

HVAC (Heating, Ventilating & Air Conditioning) /
Calefacción y Aire Acondicionado 2-14

Electrical / Eléctrico 2-16
General Terms / Términos Generales 2-16
Service Drop / Servicio Drop 2-17
Receptacle / Receptáculo 2-18
Light Fixture / Accesorios/Luces 2-18

Masonry / Albañilería 2-18
General Concrete Terms / Terminos Generales del Concreto 2-19

Framing / Enmarcado 2-21
Floor Framing / Enmarcado de Piso 2-21
Wall Framing / Enmarcado de Pared 2-21
Roof Framing / Enmarcado de Techo 2-22

Stair Parts / Partes de los Escalones 2-23

PRONOUNCING SPANISH	
Vowles	
A is pronounced	AH as in far
E is pronounced	EH as in mend
I is pronounced	EE as in feet
O is pronounced	OH as in only
U is pronounced	OO as in pool
Diphthongs (double vowels):	
io is pronounced	EEOH as in video
ie is pronounced	yeh as in yes
iu is pronounced	EEW as in few
ua is pronounced	WAH as in wallet
ue is pronounced	WEH as in well
Consonants	
h is always silent	
j is pronounced as an *h*	
ll is pronounced *y* as in yell	
ñ is pronounced *nya*, as in canyon	
ny is pronounced *nya* as in onion	
d is voiced	
t and **p** are soft	
r is soft	
rr is rolled	
ch is always as in Church	
Plurals:	
el changes to **los**	
la changes to **las**	
s or **es** are added to the end of the word.	

GENERAL COMMUNICATION / COMUNICACIÓN GENERAL

INTRODUCTION / INTRODUCCIÓN

Hello!	¡Hola!
How are you?	¿Cómo está?
It is good to meet you.	Es bueno conocerle.
Good morning.	Buenos dias.
Good afternoon.	Buenas tardes.
Good evening.	Buenas noches.
Good-bye.	Adiós.
Do you speak English?	¿Habla inglés?
What is your name?	¿Cómo se llama?/Cuál es su nombre?
My name is _____.	Me llamo _____.
	Mi nombre es _____.
please	por favor
thank you	gracias
you're welcome	de nada
yes	sí
no	no

GETTING ACQUAINTED / FAMILIARIZANDOSE O CONOCIENDOSE

How old are you?	¿Cuántos años tiene Ud.?
Are you married?	¿Es Ud. casado(a)?
How many children do you have?	¿Cuántos niños tiene?
Where do you live?	¿Dónde vive Ud.?
What's your address?	¿Cuál es su dirección?
What's your telephone number?	¿Cuál es su número de teléfono?

THE BASICS / LO BÁSICO

MONTHS / MESES

January	enero
February	febrero
March	marzo
April	abril
May	mayo
June	junio
July	julio
August	agosto
September	septiembre
October	octubre
November	noviembre
December	diciembre

DAYS OF THE WEEK / DIAS DE LA SEMANA

Sunday	domingo
Monday	lunes
Tuesday	martes
Wednesday	miércoles
Thursday	juéves
Friday	viérnes
Saturday	sábado

TIME / TIEMPO

after	después
afternoon	tarde
always	siempre
before	antes
hour	hora
late	tarde
later	más tarde
minute	minuto
morning	la mañana
never	nunca
night/evening	noche
now	ahora
on time	a tiempo
precisely (time)	en punto (tiempo)
today	hoy
tomorrow	mañana

WEATHER / TIEMPO

Celsius/centigrade	centígrados
clear	despejado
cloudy	nublado
cold	frío
dry	seco
Fahrenheit	(Fahrenheit) grados fahrenheit
freezing	congelando
frost	helada
heat	calor
hot	caliente
ice	hielo
muddy	fangoso/lodoso
rain	lluvia
rainy	lluvioso
slick	liso
slippery	resbaloso
slushy	fangoso nieve a medio derretir
snow	nieve
storm	tormenta
sun	sol

thermostat	termostato
wet	mojado
windy	ventoso

NUMBERS / NÚMEROS	
0	cero
1	uno
2	dos
3	tres
4	cuatro
5	cinco
6	seis
7	siete
8	ocho
9	nueve
10	diez
11	once
12	doce
13	trece
14	catorce
15	quince
16	dieciséis
17	diecisiete
18	dieciocho
19	diecinueve
20	veinte
21	veintiuno
22	veintidós
23	veintitres
24	veinticuatro
25	veinticinco
26	veintiseis
27	veintisiete
28	veintiocho
29	veintinueve
30	treinta
31	treinta y uno
40	cuarenta
50	cincuenta
60	sesenta
70	setenta
80	ochenta
90	noventa
100	cien
101	ciento uno
200	doscientos
300	trescientos
400	cuatrocientos
500	quinientos
600	seiscientos
700	setecientos

800	ochocientos
900	novecientos
1,000	mil
2,000	dos mil
1,000,000	un millón
2,000,000	dos millones

COLORS / COLORES	
black	negro
blue	azul
brown	café
green	verde
navy blue	azul marino
orange	anaranjado
purple	morado
red	rojo
tan	café claro
violet	violeta
white	blanco

HIRING / CONTRATANDO

ADMINISTRATION / ADMINISTRACIÓN

I need you to fill out this:	Necesito que llene esto/ esta:
- application.	- aplicación.
- emergency information.	- información de emergencia.
- federal tax form.	- formulario de impuestos federales
- I-9 form.	- formulario I-9.
- tax form.	- formulario de imouestos.

VERIFICATION / VERIFICACIÓN	
I will need a copy of one document from List A.	Necesitaré una copia de un documento de la Lista A.
OR	O
I will need a copy of one document from List B and one from List C.	Necesitaré una copia de un documento de la Lista B y uno de la Lista C.
alien registration receipt card with photograph	tarjeta de recibo de registración extranjera con fotografía
birth certificate (original or certified copy of)	certificado/acta de nacimiento (copia certificada u original del)
certification of birth abroad	certificado/acta de nacimiento extranjero
driver's license or state ID	licencia de conducir o identificación estatal
employment authorization card (unexpired)	tarjeta de autorización de empleo vigente (no vencida)
employment authorization document (unexpired)	documento de autorización de empleo vigente (no vencido)

employment authorization document issued by the INS (unexpired)	documento de autorización de empleo emitida por el INS (no vencido)
ID issued by government	identificación emitida por el gobierno
passport	pasaporte
school ID	identificación de la escuela
Social Security card	tarjeta del Seguro Social de los Estados Unidos
temporary resident card (unexpired)	tarjeta de residente temporal vigente (no vencida)

PRIOR TO HIRING / ANTES DE CONTRATAR

I need to verify your:	Necesito verificar su_____:
- address.	- dirección.
- identification.	- identificación
- insurance.	- seguro.
- general liability insurance	- seguro general de reponsabilidad.
- references.	- referencias.
- worker's comp.	- seguro de compensación del trabajador

REPORTING TO WORK / REPORTANDO A TRABAJAR

Please be here:	Por favor este aquí:
- after lunch.	- después de almuerzo.
- before noon.	- antes del medio dia.
- early.	- temprano.
- in the afternoon.	- por la tarde.
- in the morning.	- en la mañana.
- on time.	- a tiempo.
- tomorrow.	- mañana.

CONDUCT ON JOB SITE / CONDUCTA EN EL LUGAR DE TRABAJO

Please DO NOT:	Por favor NO:
- holler at your co-workers.	- gritar a sus compañeros.
- play your music too loud.	- toque su música muy alto.
- smoke on the job.	- fume en el trabajo.
- use inappropriate language or gestures.	- use lenguaje o gestos inapropiados.

COMPLIMENTS / CUMPLIDOS

Excellent!	¡Excelente!
Fabulous!	¡Fabuloso!
Good job!	¡Buen trabajo!
Great work!	¡Gran trabajo!
I am impressed!	¡Estoy impresionado!
Very good!	¡Muy bueno!

CODE ADMINISTRATION / CODIGO DE ADMINISTRACIÓN

According to the code book, you should _____.	De acuerdo al código del ibro, usted debe _____.
We are to comply with the:	Debemos obedecer con el/la:
- article.	- artículo.
- code book.	- código del libro.
- code official.	- código oficial.
- exception.	- excepción.
- inspection.	- inspección.
- inspector.	- inspector.
- International Building Code.	- Código Internacional de Construcción.
- International Fuel Gas Code.	- Código Internacional del Gas Combustible.
- International Mechanical Code.	- Código Mecánica Internacional.
- International Plumbing Code.	- Código Internacional de Plomería.
- International Residential Code.	- Código Internacional Residencial.
- National Electrical Code.	- Código Nacional de Electricidad.
- OSHA (Occupational Safety & Health Administration).	- ASSO (Admistración de Salud y Seguridad Ocupacional).
- permit.	- permiso.
- plans.	- planes/planos.
- specifications.	- especificaciones.
- Uniform Building Code.	- Código de Construcción Uniforme.
- Uniform Plumbing Code.	- Código de Plomería Uniforme.
- Uniform Mechanical Code.	- Código de Mecánica Uniforme.
- minimum.	- mínimo.
- maximum.	- máximo.
- violation.	- violación.

CONSTRUCTION PERSONNEL / PERSONAL DE CONSTRUCCIÓN

Did the _____ show up today?	¿Se presentó _____ hoy?
Have you spoken to the _____?	¿Ha hablado con el/la _____?
It is time to call _____.	Es tiempo de llamar a _____.
You will need to speak with the _____.	Necesitará hablar con el/la _____.
You will need to wait for the _____.	Necesitará esperar por el/la _____.
I am the:	Yo soy el/la:
- apprentice.	- aprendiz.
- architect.	- arquitecto.
- boss.	- jefe.

- building inspector.	- inspector de construción.
- carpenter.	- carpintero.
- contractor.	- contratista.
- crew.	- equipo de trabajo.
- designer.	- diseñador.
- electrician.	- electricista.
- engineer.	- ingeniero.
- estimator.	- estimador.
- foreman.	- jefe.
- helper.	- ayudante.
- human resource manager.	- gerente de personal/ recursos humanos.
- inspector.	- inspector.
- insurance adjustor.	- ajustador de seguros.
- journeyman.	- oficial.
- manager.	- gerente.
- office manager.	- gerente de oficina.
- operator.	- operador/a.
- owner.	- dueño.
- plumber.	- plomero.
- project manager.	- gerente del proyecto.
- roofer.	- trabajador de techos.
- superintendent.	- superintendente.
- supervisor.	- supervisor.
- technician.	- técnico.
- trim carpenter.	- carpintero de moldura.
- worker.	- trabajador.

ADMINISTRATIVE DOCUMENTS / DOCUMENTOS ADMINISTRATIVOS

Bring me the _____.	Tráeme el/la _____.
Do you have _____?	¿Tiene _____?
Have you reviewed the _____?	¿Ya revisó el/la _____?
Make sure to have the _____ signed before you begin work.	Asegurese de tener el/la _____ firmado/a antes de que comience a trabajar.
Please sign the _____.	Por favor firme el/la _____.
The _____ should be approved first.	El/la _____ debe ser aprovado/a primero.
We will need to review the _____.	Necesitamos revisar el/la _____.
You will need _____ before you begin.	Usted necesitará _____ antes de que empiece.

architectural plans	planos arquitectónicos	
bid	oferta	
bond	bonoertificate of occupancy	certificado de ocupación

change order	cambiar el orden	
construction schedule	cronograma/ horario de construcción	
contract	contrato	
daily log	registro diario	
daily report	reporte del día	
document (noun)	documento	
(to) document (verb)	documentar	
drawing(s)	dibujo(s)	
estimate	estimación	
job report	reporte de trabajo	
lien	embargo preventivo	
note(s)	nota(s)	
notice to proceed	aviso para proceder	
performance bond	obligación de interpretación	
permit	permiso	
report	reporte	
safety policy	poliza de seguridad	
safety report	reporte de seguridad	
schedule	horario	
site plan(s)	plano(s) del sitio	
structural plan(s)	plano(s) structural(es)	

SITE WORK / LUGAR DE TRABAJO

backsite	parte trasera
backhoe	retroexcavadora
bedrock	roca de fondo
benchmark	punto de referencia
blade	hojas/cuchillas
blue stakes	estacas azules
boundary	límite
bracing	refuerso
bucket	cubo/cubeta
caution tape	cinta de precaución
cave-in	colapso/ ceder
chain	cadena
clay	arcilla/barro
clod	terrón
construction	construcción
culvert	alcantarilla
curve	curva
degree(s)	grado(s)
depth	profundidad
dirt/earth	tierra
distance	distancia

drain	desague	sewage	drenaje
drainage	drenaje	shored construction	construcción apuntalada
drill	taladro	shoring	puntales/apuntalada
dump truck	camión de volteo/ volqueta	shovel	pala
dust	polvo	shrubs	arbustos
dust control	controlar el polvo	slab	plancha/losa
earth work	terrapien	slope	inclinación
egress	salida/egreso	sod	césped/grama (chamba)
elevation	elevación	soil	tierra
embankment	embancamiento	soil engineer	ingeniero de suelos
erosion	erosión	soil type	tipo de suelo
fence	cerca	spray paint	pintura en aerosol
fill	relleno	stability	estabilidad
flag stakes	estacas de bandera	steam	vapor
flagger	persona que da señales	stone	piedra
foresite	parte frontal	storm drain	drenaje para tormentas
grade	grado	string line	línea de hilo
grader	máquina niveladora	survey	medición (terreno)
gravel	grava	surveyor	topógrafo
grid lines	líneas de rejilla	swing	oscilación
hazardous communications	comunicaciones peligrosas	tamper	manipular
hole	agujero	templates	plantillas
hub	eje	theodolite	teodolito
hydrant	hidrante	topsoil	tierra negra/capa superficial de tierra
incline	declive/inclinación	topographic map	mapa topográfico
landscaper/gardener	jardinero	track hoe	excavadora
limbs	rama	traffic	tráfico
limestone	piedra caliza/para afilar	transit	tránsito
lot	terreno/lote	trash	basura
manhole	boca de acceso/alcantarilla	trees	arboles
moisture	humedad	tripod	trípode
mud	lodo	tunnel	túnel
paint mark	pintura de marcar/marca de pintura	unbalanced fill	relleno sin balance
pile	montón	underground lines	líneas subterráneas
pipe	tubo	unshored construction	construcción no apuntalada
pipeline	tubería	unstable ground	terreno inestable
pit	hoyo/pozo	utility lines	líneas de servicios públicos
plot	trazar	water table	mesa de agua
power lines	líneas de electricidad	water truck	tanquero/pipa de agua
property	propiedad	waterline	línea de agua
property boundary line	línea de límite de la propiedad	weeds	malas hierbas/monte
		winch	torno
public utilities	servicios públicos	wood stake	estaca de madera
quicksand	arena lijera/rapida		
rock	roca/piedra		
rubbish	desperdicios		
rubble	escombro		
runoff	agua de desague		
sand	arena		
sandstone	arenisca		
septic tank	fosa séptica		

TOOLS & EQUIPMENT / HERRAMIENTAS Y EQUIPO	
Do you have a/an _____?	¿Tiene un/una _____?
Please bring me a/an _____.	Por favor traigame un/una _____.
The job will require a/an _____.	El trabajo requerirá un/una _____.

This is a list of the tools you/ we will need.	Esta es una lista de las herramientas que necesitará/ necesitaremos.
This is my _____. Make sure it is returned to me.	Este es mi _____. Asegúrese que sea devuelto/a.
We need a/an _____.	Necesitamos un/una _____.
Where is the _____?	¿Dónde está el/la/los/las _____?

HAND TOOLS / HERRAMIENTAS MANUALES

adjustable wrench	llave francesa/ajustable
allen wrench	llave allen
ball-peen hammer	martillo de bola
bar	barra
basin wrench	llave pico de ganso
blade	hojas de lámina (para cortar)
bucket	cubeta
carpenter's apron	delantal de carpintero
carpenter's square	escuadra de carpintero
caulking gun	pistola enmasilladora (silicon)
C-clamp	abrazadora/prensa en "C"
chain	cadena
chalk line	línea de yeso
channel-lock pliers	tenazas de cerradura de canal
claw hammer	martillo con pinza curva
comealong	mordaza tiradora de alambre/ cabrestante
crescent wrench	llave para tuercas crecent
cutter	cortadora
file	lima
flashlight	linterna/lámpara
flat head	cabeza/hoja plana (destornillador)
framing square	escuadra para enmarcación
goggles	anteojos/gafas de seguridad
hammer	martillo
handsaw	serrucho de mano
hook	gancho
knife	cuchillo
level	nivel
mallet	mazo
metal scribe	trazador de metal
ladder	escalera
laser level	nivel de laser
paintbrush	brocha para pintar
Phillips screwdriver	destornillador/desarmador estrella (Philips)
plane	cepillo
pliers	alicates/tenazas/playo
plumb bob	plomada/plomo

plumb line	linea de plomada
putty knife	espátula para masilla
roller	rolo/rodillo
safety glasses	anteojos/gafas de seguridad
saw	serrucho
sawhorse (steel)	caballete/burro (acero)
screwdriver	destornillador/desarmador
sheet-metal shears	tijeras para metal
square	escuadra
soldering torch	soplete/soldadura de antorcha
stapler	engrapadora/grapadora
staple gun	engrapadora/ grapadora automática
string line	línea de hilo/cuerda
stud finder	detector de barrotes
tape measure	cinta para medir
tin snips	tijeretas de estaño
torque wrench	llave dinamométrica
tripod	trípode
tweezers	pinzas
vise	tornillo de banco
vise-grip pliers	alicates/tenazas de presión
wire brush	cepillo de alambre

POWER TOOLS / HERRAMIENTAS ELÉCTRICAS

air compressor	compressor de aire
chainsaw	sierra de cadena
chop saw	sierra tajadero
circular saw	sierra circular
circular saw blade	hoja de sierra circular
compound miter saw	sierra de corte angular
cut-off saw	sierra para cortar
drill	taladro
drill bit	broca para taladro
electric drill	taladro eléctrico
generator	generador
grinder	afilador/esmerilado
hammer drill	taladromartillo
impact wrench	llave de impacto
jackhammer	martillo perforador
jigsaw	sierra de vaivén
miter saw	sierra angular
miter box	caja de corte a angulos (inglete)
nail gun	pistola para clavos
powder nailer	pístola de cartuchos
radial arm saw	sierra de brazo radial
reciprocating saw	sierra oscilatoria
right-angle drill	taladro de ángulo recto
router	fresadora/contorneador/ encaminadora
sander	lijadora

saw	sierra
screw gun	pistola de tornillos
table saw	sierra de mesa
torch	antorcha

LANDSCAPE & CLEANING / JARDINERÍA Y LIMPIEZA	
ax	hacha
blower	sopladora
broom	escoba
brush	pincel/brocha/cepillo
bucket	cubeta/balde
cart	carreta/carretón
chemical	químico
Dumpster	contenedor de escombros
dust mop	trapeador para polvo
dustpan	recogedor de polvo
garden hose	manguera de jardín
gloves	guantes
hoe	azadón
hose	manguera
mop	trapeador
mop bucket	cubeta para trapeador
pick	pico
push broom	escobón
rags	trapos
rake	rastrillo
shovel	pala
sink	fregadero
sledgehammer	mazo
soap	jabón
spray bottle	bote para rociar
squeegee	enjugador
trash	basura
(to) vacuum (*verb*)	aspirer
vacuum cleaner	aspiradora
worm-drive saw	sierra circular con tornillo sin fin
wheelbarrow	carretilla de rueda

MOTORIZED / MOTORIZADO	
backhoe	retroexcavadora
bulldozer	Niveladora
crane	grúa
dump truck	volqueta
flat-bed truck	camión de tarima
scissor lift	plataforma hidráulica
truck	camión
front-end loader	cargador de final delantero
forklift	montacargas
mixer	mezcladora
pump	bomba

AREAS OF A HOUSE OR BUILDING / AREAS DE LA CASA O DEL EDIFICIO	
aisle	pasillo
attic	atico/desván
atrium	atrio
basement	sótano
bathroom	baño
bedroom	habitación/dormitorio
boiler room	cuarto del calentador de agua
bonus/extra room	cuarto adicional
breakfast room	desayunador
butler's pantry	despensa
canopy	toldo
ceiling	cielo
closet	armario
corridor	pasillo
crawl space	debajo de la casa
deck	balcón posterior
dining room	comedor
driveway	camino a la entrada (casa)
dwelling	vivienda
dwelling unit	unidad de vivienda
elevator	elevador
entry	entrada
exit	salida
floor	piso
great room	sala grande
hallway	vestíbulo/pasillo
hall closet	armario de pasillo
kitchen	cocina
living room	sala
mezzanine	entresuelo/ mezzanine
mud room	cuarto de barro
pantry	despensa
penthouse	atico
porch	pórtico
roof	techo
room	cuarto
sidewalk	acera
stairs	escaleras/gradas
storage room	cuarto de almacenamiento
vestibule	vestíbulo
walk-in closet	armario grande
wall	pared

SAFETY / SEGURIDAD	
DISTRESS SITUATIONS / SITUACIONES DE PELIGRO	
Are you hurt?	¿Está herido?
Call for help.	Llame por ayuda.
Don't move.	No se mueva.

Do you use drugs?	¿Usa drogas o narcóticos?
Do you take any medication?	¿Está tomando algún medicamento?
Go for help.	Vaya por ayuda.
Get the first-aid kit.	Traígame el botiquín de primeros auxilios.
I called 9-1-1.	Llamé al 9-1-1.

INJURY TERMS / TERMINOS PARA LESIONES

We need _____.	Necesitamos _____.
He/she will need _____.	El/ella necesitará _____.
ambulance	ambulancia
antibiotic	antibiótico
bandage	venda/vendaje
dehydrated	deshidratado
defibrillator	defibrilador (máquina para dar choques eléctricos)
emergency	emergencia
CPR (cardiopulmonary resuscitation)	RCP (respiración cardio-pulmonar)
heatstroke	insolación
hospital	hospital
medication/medicine	medicamento/medicina
poison	veneno
splint	tablilla
sterile	estéril
tetanus shot	inyección contra el tetanus/tétano

FIRST-AID KIT / BOTIQUÍN DE PRIMEROS AUXILIOS

The first-aid kit is:	El botiquín está _____:
- in the office.	- en la oficina.
- in the trailer.	- en la traila.
- in the vehicle.	- en el carro.
Contents of the first-aid kit should include:	El contenido del botiquín debe incluir:
- adhesive tape	- cinta adeshiva
- antacid	- antiácido
- aspirin	- aspirina
- baking soda	- bicarbonato de sodio
- cotton balls/swabs	- bolas de algodón
- decongestant	- descongestionante
- face mask for CPR	- máscara para RCP
- first-aid guide	- guía de primeros auxilios
- flashlight	- lámpara/linterna
- gauze pads	- gazas
- hot-water bottle	- botella de agua caliente
- household ammonia	- amonia de uso casero
- hydrocortisone cream	- hidrocortizona en crema
- hydrogen peroxide	- peróxido hidrogeno
- ice bag/ice pack	- bolsa de hielo

- insect repellent	- repelente de insectos
- latex gloves	- guantes de latex
- matches	- fósforos/cerillos
- meat tenderizer	- hablandador de carne
- moleskin	- piel de topo
- needles	- agujas
- safety pins	- ganchitos de seguridad
- salt	- sal
- scissors	- tijeras
- soap	- jabón
- splint	- tablilla
- thermometer	- termómetro
- tweezers	- pinzas
- waterproof tape	- cinta a prueba de agua

INJURED PERSON / PERSONA LESIONADA

Can you move your _____?	¿Puede mover su ____?
Does your _____ hurt?	¿Le duele su _____?
Are you bleeding?	¿Está sangrando?
ankle	tobillo
arm	brazo
back	espalda
bicep	bícep
bleeding	sangrando
blood	sangre
butt	gluteos (nalga)
chest	pecho
chin	barbilla
ear	oreja
elbow	codo
eye	ojo
finger	dedo de la mano
foot/feet	pie/pies
groin	ingle
hand	mano
head	cabeza
heart	corazón
hip	cadera
knee	rodilla
leg	pierna
mouth	boca
neck	cuello
nose	naríz
shoulder	hombro
stomach	estómago
tooth/teeth	diente/dientes
thigh	muslo
thumb	dedo pulgar
toe	dedo del pie
wrist	muñeca de la mano

HAZARDS / RIEZGOS	
asbestos	asbesto
fire	fuego
gasoline	gasolina
lockout	cierre/bloqueo
MSDS (Material Safety Data Sheet)	HDMS (Hoja de Datos de Material Seguro)
paint thinners	diluyentes de pintura
sliver	astilla

SAFETY WARNINGS / ADVERTENCIAS DE SEGURIDAD	
Be careful!	¡Tenga cuidado!
Watch out!	¡Ponga atención!
Keep out.	Manténgase fuera/alejado.
Please follow all safety rules.	Por favor siga todas las reglas de seguridad.
Report all accidents immediately.	Reporte todos los accidentes de inmediato.

SAFETY INSTRUCTIONS / INSTRUCCIONES DE SEGURIDAD	
Be careful, you are working with material that is:	Tenga cuidado usted está trabajando con materiales que son/es:
- caustic.	- cáustico.
- combustible.	- combustible.
- explosive.	- explosivo.
- flammable.	- inflamable.
- hazardous.	- peligroso.

PERSONAL PROTECTIVE EQUIPMENT / EQUIPO DE PROTECCIÓN PERSONAL	
Hard hats and safety glasses are required at all times.	El casco y las gafas/lentes de seguridad se requieren a todo el tiempo.
You must always wear your _____.	Siempre debe llevar puestos sus _____.
You will always need _____ for this job.	Siempre necesitará _____ para este trabajo.
dust mask	mascarilla para polvo
earplugs	tapones para los oídos
eye protection	protección para (los ojos)
fall harness	caída de arnes
gloves	guantes
goggles	anteojos/gafas
hard hat	casco
head protection	protección para la cabeza
lanyard	acollador
leather gloves	guantes de cuero
respirator	respirador
rubber boots	botas de goma/hule/caucho

rubber gloves	guantes de goma/hule/caucho
steel-toe boots	botas con protección de metal en la punta
vests	chalecos
welding apron	delantal de soldar

SIGNS / SEÑALES	
Pay attention to all warning signs.	Ponga atención a todas las señales de advertencia.
Make sure this sign is posted at all times.	Asegúrese que ésta señal esté puesta todo el tiempo.
Authorized Personnel Only	Solo Personal Autorizado
Construction Area	Area de Construcción
Construction Entrance	Entrada a la Construcción
Construction in Progress	Construcción en Progreso
Construction Zone	Zona de Construcción
Confined Space	Espacio Confinado/limitado
Danger!	¡Peligro!
Deep Excavation	Excavación Profunda
Do Not Enter	No Entrar
Enter with Permit Only	Entrar Solo con Permiso
Gloves Required	Guantes Requeridos
Goggles Required	Anteojos/Lentes Requeridos
Hard Hat Area	Area de Cascos
Hard Hats Required	Cascos Requerido
No Trespassing	No Traspasar
Open Hole	Agujero Abierto
Open Pit	Hoyo Abierto
Open Trench	Trinchera Abierta
People Working	Gente Trabajando
Personal Protective Equipment Required	Equipo de Protección Personal Requerido
Restricted Area	Area Restringida
Steel-Toe Shoes Required	Zapatos con Punta de Metal Requerido
Warning!	¡Advertencia!
Watch for Moving Equipment	Cuidado con el Equipo en Movimiento
Watch for Falling Debris	Cuidado Escombros Cayendo

DRYWALL / TABLA ROCA O TABLA DE YESO	
adhesive	pegamento
all-purpose compound	masilla para todo uso
backing/backer board	tablilla de yeso
base coat	cobertura/aplicación base
baseboard	zócalos/moldura de base
beveled edge	angulo biselado
blister	burbuja
bullnose	canto biselado
butt joint	junta recta

chalk line	línea de yeso		
clickers	clicker (clic)		
control joints	uniones de control		
corner bead	protector de esquina		
dimple	hoyuelo/hoyito		
drywall primer	pintura base para tabla roca		
drywall	tabla roca/tabla de yeso		
EIFS (Exterior Insulation Finish System)	Sistema de Acabado de Aislamiento/Insulación Exterior		
field (bolt)	atornillar/empernar		
fire taping	cinta anti-fuego		
firewall (drywall)	pared anti-fuego		
furring	enrasado		
greenboard	tabla verde		
gypsum	yeso		
insulation	insulación/aislamiento		
knife	cuchillo/espátula		
knockdown	demoler/tumbar		
mud	masilla		
multilayer	capas multiples		
nail-pop	clavo salido/saltado		
orange peel (texture)	piel de naranja (textura)		
plumb	plomada		
popcorn (texture)	textura como palomitas de maíz		
primer	pintura base		
PVA (polyvinyl acetate)	acetato de polivinilo		
repair	reparar		
ring shank	fuste corrugado		
rip	razgar/tirar		
ripper	chísel		
roto zip	rotozip		
router	encaminador/broca or direccionador		
R-value (thermal resistance)	resistencia térmica		
sand swirl	lijar en círculos		
score	resultado		
seam	costura/unión		
self-adhesive joint tape	cinta de unión autoadhesiva		
sheetrock (drywall)	tabla roca		
skim coat	aplicar masilla ligeramente		
skip trowel	espátula con cuchilla		
slapbrush	brocha empalmada		
smooth wall	pared suave/lisa		
soffit	sofito		
spotting	manchar		
stud	montantes		
T-square	escuadra T		
texture	textura		
topping joint compound	masilla para uniones		
trowel	espátula/paleta		

PAINT / PINTURA	
abatement	remoción/disminución (pollution - contaminación)
adhesion	adhesión
acrylic	acrílico
angle sash	angulo inclinado
aerosol	aerosol
air dry	aire seco
air cure	aire muy seco
alkali	alcali
aluminum paint	pintura de aluminio
amide	entre/amida
anchoring	anclamiento
anti-corrosive paint	pintura anti-corrosiva
antique finish	acabado antiguo
back primed	imprimador de base
benzene	benceno
benzine	bencina
binder	unión para junturas
bleaching	descolorante
bleeding	sangrante
blistering	abrasador
blushing	ruborizante
body	cuerpo
breathe	respirar/soplar
bristle	cerda
brushability	cepillable
brush comb	peine para brochas
brush marks	marcas de brocha
brush out	brochar hacia afuera
bubbles	burbujas
calcimine	pintura de cal
camel hair	pelo de camello
catalyst	catalizador
chalking	caliza/yeso / tiza
china bristle	cerda de china
chroma	intensidad de color
clear coating	capa clara / capa transparente
coverage	cobertura
crazing	moda
creosote	chapote
custom color	color hecho/hacer
cutting in	recorte hacia dentro
denatured alcohol	alcohol desnaturalizado
drywall compound	masilla de tabla roca
durability	durabilidad
dye	tinte
eggshell	cascara de huevo
fading	descoloración
feather sanding	lijada lijera
filler	relleno
film	capa

flaking	descascararse
flat	mate/sin brillo
flat oil	aceite mate/sin brillo
floating	flotante
fungicide	fungicida
glaze	glaseado/cristalizado
glazing compound	masilla glaseada
gloss	brilloso
hardness	dureza
holiday	temporada
hot-dog roller	rodillo tipo "hot dog"
inhibitor	inhibidor
interior	interior
intermediate coat	capa intermedia
lacquer	laca
lambswool	lana de cordero
lap	vuelta
latex	latex
leveling	nivelando
linseed oil	aceite de semilla de linaza
masking	aislante
masking tape	cinta aislante
matte	mate
mildew resistance	resistencia al moho
mildewcide	anticorrosivo
mineral spirits	destilados de petróleo
nap	pelusa/ lanilla
nylon	nilón/nailon
opacity	opacidad
opaque coating	capa opaco/a
orange peel	cáscara de naranja (textura)
polyester	poliester
paint	pintura
paint remover	removedor de pintura
paint spot	mancha de pintura
painter's glove	guante de pintores
peeling	pelando
pigments	pigmentos
pinhole	hoyo para clavo
polyurethane	poliuretano
prime coat	capa principal
primer	imprimación / tapaporos
putty	masilla
putty knife	espátula/cuchilla para masilla
red oxide	oxido rojo
resin	resina
roller	rodillo/rolo
run	chorrear/correr
sags	combas
sand finish	acabado arenoso
satin	satin
semi-gloss	semi-brilloso

semi-transparent	semi-transparente
settling	estableciendo
shellac	laca
silicone	silicón
solvent	solvente
spackling compound	pasta de masilla
spatter	salpicar
spot priming	imprimación de manchas
spraying	rociar
spreading rate	cantidad de aplicación
stain	tinte
streaking	rayado
strip	tira
substrate	substrato
tack rag	trapo con pega (para limpiar pelusa)
tacky	pegagoso
texture	textura
texture paint	textura de la pintura
thinner	disolvente
tint base	tinte base
toner	tóner
touch up	retocar
turpentine	trementina
undercoat	primera mano
varnish	barniz
varnish stain	tinte de barniz
viscosity	viscocidad
washability	lavabilidad
water spotting	manchas de agua
weathering	desgaste
wood filler	relleno para madera
wire brush	cepillo de alambre
yellowing	amarillento

ROOF COVERING / REVESTIMIENTO DE TECHO	
asphalt	asfalto
asphalt shingle	teja de asfalto
caulk	calafatear
curb	guarnición
dead load	carga muerta
deck/decking	cubierta
dormer	buharda
eave	alero
felt	gamuza
flashing	cubrejuntas/tapajuntas
flat roof	techo plano
gable	hastial
gable roof	techo a dos aguas
gutter	canalón
hatch	compuerta
hip	lima

hip roof	techo a cuatro aguas
mansard roof	techo mansardara /a boardillado
metal deck	plataforma metálica
overhang	voladizo/vuelo/alero
rafter	viga
ridge	cumbrera
ridge board	tabla de cumbrera
roof	techo
roof covering	revestimiento de techo
roof deck	cubierta de techo
roof drain	desague de techo
roof sheathing	entarimado de teja
roof tile	teja
roofer	techero
roofing felt	teja asfáltica/felpa
roofing square	cuadrado de teja
sealant	sellador
shingle	teja
skylight	tragaluz/claraboya
slate shingle	teja de pizarro
tar	alquitran/brea/chapopote
tar paper	papel de brea
waterproofing	impermeable
wood shake	teja de madera/ripia
wood shingle	teja de madera

PLUMBING / PLOMERÍA

GENERAL TERMS / TÉRMINOS GENERALES

back flow	contraflujo
backing	soporte
ball cock	válvula de flotador
ball valve	llave de flujo
bleeder valve	válvula de purga
bathroom	cuarto de baño
bathroom sink	lavabo de baño
bathtub	bañera
bracket	brazo
branch	ramal
brass	bronce
braze	soldar
brazing alloy	aleación para soldar
brazing flux	fundente para soldar
chase	canaletas
check valve	válvula de contraflujo
cistern	cisterna
cleanout	registro
cold supply	proveedor frío
cut-off valve	válvula de cierre
diameter	diámetro
elbow	codo

faucet	llave de agua
grease interceptor	interceptor de grasas
grease trap	trampa de grasas
hose bibb	grifo de manguera
hot supply	suministro de agua caliente
hub valve	válvula de cubo
hydrant	hidrante
key valve	válvula de llave
pipe	tubo
plumber	plomero
plunger	destapacaños/desatascador
plumbing	plomeria
plumbing appliance	muebles del sanitario
propane	propano
pump	bomba
radius	radio
relieve valve	válvula auxiliar
septic tank	fosa séptica
sill cock	grifo de manguera
shower stall	ducha/regadera
shut-off valve	válvula de cierre
soil pipe	tubo de aguas negras
soil stack	(tubo) vertical de aguas negras
spigot	llave/grifo/canilla
sprinkler	rociador
sprinkler system	sistema de rociadores
stack	tuberia vertical
stack vent	respiradero vertical
standpipe	columna hidrante
standpipe system	sistema de columna hidrante
sump	sumidero
sump pump	bomba de sumidero
sump vent	respiradero de sumidero
thread	rosca
tee	"T"
tempered water	agua tibia
tempering valve	mesclador (en línea)
template(s)	plantilla(s)
toilet	inodoro/sanitario
torch	antorcha
trap	trampa
trap seal	sello de trampa
valve	válvula
vent shaft	recinto de ventilación
vent stack	respirador vertical/ventila principal
vent system	sistema de ventilación
venting system	sistema de ventilación
washer and dryer	lavadora y secadora
wax seal	empaque de cera
well (water)	aljibe/pozo de agua

SPANISH

TOILET / INODORO

adjust clip	broche de ajuste
closed end	terminación cerrada
compression fitting	accesorios de compresión
coupling nut	tuerca de cople
fill line	línea de llenar (agua)
fill valve	válvula de llenar (agua)
filling flow	fluido de llenamiento
flange bolt	cerrojo de reborde
flapper	tapón/descarga del inodoro
float arm	brazo flotante
float cup	taza flotante
flush handle	manija para inodoro
height adjust	adaptador de altura
horn	cuerno
lift chain	cadena para levantar
link	conexión
overflow tube	tubo para rebalsamiento
rim hole	agujero de borde
shut-off valve	válvula de cierre
siphon jet	chorro por sifón
spud nut	tuerca para base
spud washer	arandela para base
supply tube	tubo de suministro
tank	tanque
toilet range	ajustador para inodoro
trap weir	presa de trampa
waste line	línea de desecho
wax ring	anillo de cera

FAUCETS / GRIFOS O LLAVES

adjusting ring	anillo ajustante
alignment pin	perno de alineamiento
ball	bola
base	base
body	cuerpo/masa
cam	leva
cap	tapa/tapón
cylinder	cilindro
escutcheon	escudo
faucet body	masa del grifo/llave
handle	manguillo
handle screw	tornillo de manija
inlet seal	sello de entrada
notch	muesca
O-ring	anillo "O"
outlet seal	sello de salida
packing	embalaje
packing nut	tuerca de embalaje
plastic retaining nut	tuerca de retención plástica
replaceable cartridge	cartucho reemplazable

retaining clip	clip de retención
retaining clip slot	clip de retención en ranura
rubber cam seal	sello de leva de goma / sello de leva de caucho
rubber inlet seal	sello de entrada de goma
screw	tornillo
set screw	juego de tornillos
spindle	eje
spout	canilla
spring	resorte
stem assembly	ensamble de tallo
stem screw	tornillo de tallo
stem washer	arandela de tallo
tab	pestaña/etiqueta

TUB/SHOWER CONTROLS / TINA/CONTROLES DE LA DUCHA/REGADERA

cartridge	cartucho
cold supply	suministro de agua fría
control handle	manija/manecilla de control
diverter	desviador
escutcheon	escudo
friction ring	anillo de fricción
handle	manija/manguillo
hot supply	suministro de agua caliente
inlet seal(s)	sello(s) de entrada
outlet seal(s)	sello(s) de salida
shower control	control de ducha/regadera
stem	tallo
tub spout	canilla para tina
valve seat	asiento de válvula
washer	arandela

HVAC (HEATING, VENTILATING & AIR CONDITIONING) / CALEFACCIÓN Y AIRE ACONDICIONADO

absolute	absoluto
absorbent	absorvente
absorption	absorción
active cooling	refrigeración activa
accumulator	acumulador
air balance	balance de aire
air conditioner	aire acondicionador
air conditioning	aire acondicionado
air handler	tratante de aire
air infiltration	infiltración de aire
air vent	abertura de aire
ambient temperature	temperatura ambiente
ammeter	amperímetro
ampere	amperio
amplitude	amplitud
analog	análogo
anode	anodo

aspiration	aspiración	evacuation	evacuación
atmospheric condenser	condensador atmosférico	evaporation	evaporación
back pressure	presión trasera	exfiltration	exfiltración
baffle	separadores	Fahrenheit	grados (Fahrenheit)
balanced pressure	presión balanceada	fan	ventilador
barometer	barómetro	filter	filtro
bleed valve	sangrar válvula	fire damper	apagador de fuego
blend	mezclar	flammability	inflamabilidad
blower	soplador	flapper valve	válvula direccional
boiler	calentador de agua	flexible duct	conducto flexible
boiling	ebullición / nervir	float valve	válvula flotante
booster	elevador de voltaje	flow meter	contador de flujo
BTU (British thermal unit)	(UTB) unidad térmica Británica	flue	conducto de humo
		fluid	fluido
burner	quemador	flush	unión
bypass	bypass	flux	fundente
calibrate	calibrar	force	fuerza
capacity	capacidad	freezing point	punto de congelación
carbon dioxide	dióxido de carbono	Freon	Refrigerante
carbon monoxide	monóxido de carbono	frost-free	libre de congelación
Celsius/centigrade	centígrados	fuse	fusible
CFM (cubic feet per minute)	PCM (pies cúbicos por minuto)	gas	gas
		gas valve	válvula de gas
charge	carga	gas regulator	regulador de gas
check valve	válvula de probar	gasket	junta
chiller	enfriador	gauge	medida / calibrador
chill factor	factor frío	glide	deslizar
chimney	chimenea	glycerol	glicerol
circuit	circuito	gravity cooling	gravedad de enfriamiento
circuit breaker	interruptor de circuito	grille	reja
cold	frío	grommet	ojal plástico o de metal
comfort zone	area de confort	ground coil	bobina de tierra
compressor	compresor	ground loop	cable de tierra (vuelta)
condense	condensar	ground wire	cable de tierra
condenser	condensador	hanger	suspensión
conductivity	conductividad	head	cabeza/principal
damper	apagador	head pressure	presión principal
defrost	descongelar	head velocity	velocidad principal
defrost cycle	ciclo de descongelar	heat	calor
dehumidify	deshumedecer	heat content	contenido de calor
dehumidifier	deshumidificador	heat intensity	intensidad de calor
dehydrator	deshidratador	heat lag	retraso de calor
dew	rocio	heat load	carga de calor
dew point	punto de rocio	heat pump	bomba de calor
draft	infiltración de aire	heat source	fuente de calor
drip pan	cazuela de goteo	heat transfer	transferencia de calor
duct	ducto	hertz (Hz)	hercio
EER (Energy Efficiency Ratio)	Proporción de Energía Eficiente	high-vacuum pump	bomba neumática
		humidifier	humidificador
electric heating	calefacción eléctrica	humidity	humedad
electrolysis	electrólisis	HVAC (heating, ventilating & air conditioning)	calefacción y aire acondicionado
environment	ambiente		

hydraulic	hidráulico
idler	direccional (polea)
ignition transformer	transformador de ignición
impedance	impedancia
impeller	asta
induction motor	motor de inducción
inductance	inductancia
infiltration	infiltración
isobutane	isobutane
isothermal	isotérmico
jet cooling system	sistema de refrigeración reactor
joule	julio (medida)
junction box	caja de conexiones
Kata thermometer	termómetro Kata
Kelvin scale	escala Kelvin
kilowatt	kilovatio
latent heat	calor latente
leak detector	detector de agujero
liquid line	línea de líquido
load	carga
low side	lado bajo
mass	masa
metering device	medición de dispositivo
motor	motor
needle valve	válvula de aguja
nipple	rosca
oil trap	trampa de aceite
open compressor	compresor abierto
open loop	vuelta abierta
oversized evaporator	evaporador (muy grande)
photocell	fotocélula
pressure	presión
pyrometer	pirómetro
purge	purgar
primary air	aire primario
radiant floor heating	calefacción de piso radiante
radiation	radiación
refrigeration	refrigeración
relative humidity	humedad relativa
relay	relevador
relief valve	válvula de relevo
resistance	resistencia
reversing heat pump	bomba de inversión de calor
schematic	esquemático
secondary air	aire secundario
sludge	fango/lodo
split system	sistema separado
sub-cooled	refrigeración subterranea
suction line	línea de succión
sweating	soldar
temperature	temperatura

ternary	ternario/trino
thermistor	termistor
thermometer	termómetro
thermostat	termostato
ton	tonelada
TX transducer	transductor de transmisión
unit system	unidad de sistema
vacuum	aspirar
vapor pressure	presión de vapor
volt	voltio
voltmeter	medidor de voltaje
viscosity	viscocidad
watt	vatio
water-to-water	agua a agua
wet compression	compresión húmeda
xylene	xileno

ELECTRICAL / ELÉCTRICO

GENERAL TERMS / TÉRMINOS GENERALES

AC current	corriente CA
access	acceso
access cover	tapa de acceso
alarm	alarma
aluminum	aluminio
amperes	amperias
amperage	amperaje
ballust	lastre
battery	pila
bonding conductor	cable de enloace
bonding jumper	borne de enlace
buried cable	cable enterrado
bury	enterrar
cable tray	bandeja de portacables
cabinets	gabinetes
cartridge fuse	fusible de cartucho
chase	canaletas
circuit	circuito
circuit breaker	interruptor de circuito
circuit-breaker panel	caja de cortacircuito
conduit	conducto/canal
connection	conexión/unión
connector	conector
copper	cobre
coupling	acoplamiento/cople
DC current	corriente CC
diameter	diámetro
disconnect switch	desconectar interruptor
doorbell	timbre
double-pole breaker	interruptor automático bipolar
electrician	electricista

electricity	electricidad
elbow	codo
electrical fixture	accesorios eléctricos (luces)
electrical outlet	enchufe/tomacorriente
enclosure	cerramiento
exhaust	escape/extracción
exhaust fan	ventilador de extracción
extension cord	cable de extensión
fan	ventilador
feeder cable	cable de alimentación
fire alarm	alarma de incendio
fire-proofing	ignifugación/a prueba de fuego
fitting	accesorio
fixture	artefacto/accesorio (luces)
flexible conduit	conducto portacables flexible
floodlight	iluminación
fluorescent	fluorescente
frequency	frecuencia
fuse	fusible
fuse box	caja de fusibles
generator	generador
ground bar	barra de tierra
ground rod	barra/vara de tierra
ground connection	conexión en tierra
ground fault circuit interrupter	circuito interruptor de seguridad a tierra
ground wire	alambre de tierra
hangers	ganchos
hot bus bar	bandeja de carga
impedance	impedencia
insulation	aislamiento/insulación
junction box	caja de conexiones
knockout	agujero ciego
lamp	lámpara
lightbulb	foco/ bombilla
lights	luces
live wires	alambres vivos
lockout/tagout	cerrar/etiquetar
loop	lazadas/vuelta
main	principal/matriz
main breaker	interruptor automático principal
main power cable	cable principal eléctrico
manhole	boca de accesso / alcantarilla
meter	metro/medidor
motor	motor
neutral	neutro/neutral
neutral bar	bandeja neutral
neutral conductor	cable/conductor neutral
offset	desplazamiento/desvío

open air	aire libre
outlet box	caja de enchufe/ tomacorriente
parallel	paralelo
plastic insulator	aislante plástico
phase	fase
pipe	pipa/línea
plenum	pieno/cámara de distribución de aire
plug fuse	fusible de rosca
power	potencia/energía
power lines	líneas de energía
power strip	corta picos
power supply	fuente/suministro de energía
raceway	conducto eléctrico
radius	radio
range outlet	tomacorriente/enchufe para estufa
receptacle/plug	receptáculo/enchufe
resistance	resistencia
riser pipe	tubo vertical
rough-in	instalación en obra negra/ gruesa
screw connector	conector con tornillo
series	series
service entrance	servicio de entrada
single-pole breaker	interruptor automático unipolar
sleeve	camisa/manga
smoke detector	sensor/detector de humo
sparks	chispas
splice	empalme/translape/junta
switch	interruptor/apagador
switch plate	placa de interruptor
telephone	teléfono
temporary	provisional/temporario
temporary power	energía temporaria
transformer	transformador
underground lines	líneas subterráneas
vacuum breaker	interruptor de vacio
voltage	voltaje
volts	voltios
wire	alambre/cable
wire connectors	conectores de alambre/ cable/alambre conector

SERVICE DROP	SERVICIO DROP
ground rod	barra/vara a tierra
service entrance cable	servicio de cable de entrada
transformer	transformador
utility meter	medidor de utilidad
weather head	cabeza de resistencia

SPANISH

wire A/B (hot)	alambre/cable "A"/"B" (caliente)
wire B (hot)	alambre/cable "B" (caliente)
wire C (neutral)	alambre/cable "C" (neutral)

RECEPTACLE/RECEPTÁCULO	
120 volts	120 voltios
120/240 volts	120/240 voltios
15 amps	15 amperios
20 amps	20 amperios
240 volts	240 voltios
30 amps	30 amperios
3-way switch	interruptor de 3 líneas
4-way switch	interruptor de 4 líneas
50 amps	50 amperios
AFCI	AFCI
brass terminal (black or red wire)	terminal de cobre (alambre negro o rojo)
common	común
cover plate	placa covertorora
dimmer switch	regulador de luz
double-pole/single-throw	interruptor automatic bipolar/ unipolar
fault sensor	sensor de falla
firing capacitor	capacitador de incendio
GFCI (ground fault circuit interrupter)	interruptor del circuito de fallos en toma a tierra)
ground (rounded slot)	tierra (ranura dedonda)
ground fault	caida de tierra
ground terminal (green or bare)	terminal de tierra (verde o descubierto)
hot	caliente
hot (short slot)	caliente (ranura corta)
in	en/dentro
jumper tab (remove to separate)	separador de borne
light at the end of circuit	luz al final del circuito
light in middle of circuit	luz en medio del circuito
line	línea
load	carga
mounting screw	montaje de tornillo
neutral	neutro
neutral (long slot)	neutro (ranura larga)
out	fuera
pickup coil	recoger el rollo
series of receptacles	serie de receptáculos
silver terminal (white wire)	terminal de plata (alambre blanco)
single-pole/single-throw	interruptor automáticouni-polar / unipolar
smoothing capacitor	capacitador
split-circuit receptacle	receptáculo separador de circuito

split-switched receptacle	receptáculo separador de interruptor
test	prueba
traveler	viajero
variable resistor	resistencia variable

LIGHT FIXTURE / ACCESORIOS/LUCES	
adjustable crossbar	travesaño ajustable
brass screw	tornillo de cobre
bulb	foco/bombillo de luz
cable	cable
canopy	dosel
crossbar	travesaño
fixture wire threaded (cut to length)	alambre/cable de encuentro enhebrado (cortar a lo largo)
globe	globo
globe screw	tornillo de globo
gold or silver lamp cord	cuerda de lámpara color oro o plata
grounding wire	base de alámbre
hot wire	alambre/cable caliente
junction box	caja de conecciones
lock nut	tuerca de cerradura
mounting screw	tornillo de montaje
neutral wire	alambre neutro
screw collar	cuello del tornillo
securing ring	anillo de seguridad
silver screw	tornillo de plata
socket	enchufe/tomacorriente
thermal insulation	insulación termal
threaded nipple	borde enebrado
wire nut	tuerca de alambre

MASONRY / ALBAÑILERÍA	
abutment	estribo
adhesives	pegamentos
adobe brick	ladrillo de adobe
anchors	anclas
anchor bolts	cerrojos de ancla
angle iron	hierro de ángulo
arch	arco
bat	pegar/empatar/empalmar
beaded joint	junta empalmada/ enchaflanada
bearing wall	pared de cojinete soporte
bed joint	unión/junta de cama
bevel	biselada/achaflanada
bond	adherir/pegar
brick	ladrillo
brick kiln	ladrillo horneado
brick ledge	repisa de ladrillo

bull header	ladrillo aplantillado
buttering	rectángulo
cavity wall	doble pared hueca
cinder block	bloque de ceniza
composite wall	pared compuesta
concave joint	unión cóncava
concrete block	bloque de concreto
concrete masonry unit	unidad de albañilería de concreto
coping	adaptación
course	curso
cull	eliminación
cultured stone	piedra cultivada
deflection	desviación
drip	goteo
dry saw	sierra para cortar en seco
efflorescence	eflorescente
elastic	elástico
elliptical arch	arco elíptico
epoxy mortar	cemento/mezcla epoxy
expansion joint	unión de extención
extrude	saque
extruded joint	junta/unión de saque
face	cara
face brick	cara del ladrillo
flashing	tapajuntas
firebrick	ladrillo refractario
fluted block	bloque estriado
furrowing	enrasando
glass block	bloque de vidrio/cristal
grout	lechada de cemento
halfback	medio
hard brick	ladrillo solido
head joint	junta principal
header	dintel/viga de cabecera
interlock	trabar/enganchar
jack arch	arco de jack
keystone	clave
level	nivel
lintel	dintel
load bearing	porte de carga
mason	albañil
masonry	albañilería
mortar	mezcla
natural stone	piedra natural
pier	embarcadero
pilaster	pilastra
plaster	yeso
plumb	plomada
point	punto
Portland cement	cemento de Portland
quoin	piedra angular/cuña

rake joint	unión de rastrillo
ready-mix concrete	mezcla lista de concreto
rebar	barra nueva
reinforced masonry	albañilería reforzada
rock face block	bloque con cara de roca
rowlock	hilera
solid masonry wall	pared de ladrillo sólido
specifications	especificaciones
stinger	aguijón
stretcher	ensanchador
substrate	substrato
texture	textura
trowel	paleta
Tyndall stone	piedra de Tyndall
vee joint	junta "V"
veneered wall	pared chapiada
wall ties	lazos de la pared
weathered joint	juntas de resistencia
weep hole	desague
wet saw	sierra para cortar con agua
Z-ties	Z-lazos/lazos "Z"

GENERAL CONCRETE TERMS / TERMINOS GENERALES DEL CONCRETO	
acetylene	acetileno
additives	aditivos
aggregate fines	finos de agregado
aggregates	agregados
air-entraining agent	agente inclusor de aire
backup bar	varillas adicionales
bar bender	doblador de varilla
base plate	placa de base
batter boards	camilla
beam	viga
beam forms	formas de viga
blistering	abrasadero
bolt patterns	muestra de pernos
boltcutters	cortapernos
bonding agent	agente de pegamento
boots	botas
bottle	botella
bottle caps	tapas de botella
brace	jabalcón
bulkhead	tope de formo/mampara
calcium	calcio
caps	tapas
cement	cemento
chamfer	chaflán
chute	canal inclinado
circular saw	sierra circular
clamp	grapa
clay	arcilla/barro

clearance	espacio	kneepads	rodilleras
cleat	tirante de formado	lighter	encendedor
clod	gelba	lime putty	masilla de cal
column	columna	longitudinal and transverse bars	varillas longitudinales y tranversales
concrete	concreto	mat	estera
concrete broom	escoba de concreto	moist curing	curación de humedad
concrete burns	quemaduras de cemento	oxygen	oxígeno
concrete-mixing truck	camión mezclador de concreto	panel	panel/plancha
cone ties	ligaduras cono de pared	pea gravel	grava pequeña
control joint	junta de control	pencil	lápiz
corner bar	varilla de esquina	pier	estribo
corners	esquinas	pier and column forms	formas de estribo y columna
covering	recubrimiento	pilaster	pilastra
cracks	rajaduras/grietas	pinch point	punto de pellizco
cubic feet	pies cúbicos	pins	pernos/pines
cubic yards	yardas cúbicas	planks	tablones
curb and gutter	arroyo encintado	plasticizer	plasticizador
curing compound	masilla curada	plumb bob	plomo
curtains	cortinas	positioning chain	cadena de disposición
cutting torch	boquilla de corte	power trowel	llave mecánica
D-ring	anillo en "D"	prestressed concrete	hormigón prefatigado/precargado
darby	fratás	pump	bomba
dowel	pasador de varilla	pump truck	bomba de concreto
driveway	via de acceso	quicklime	cal preparada
dust	polvo	rebar	barra de refuerzo/varilla
edge forms	formas de borde	rebar caps	tapas de varilla
embed plate	placa embutida	rebar shears	tijeras de varilla
expansion bolt	perno de expansión	regulator valves	válvulas de regulación
expansion joint	junta de expansión	reinforcement	refuerzo
extension cord	cordón de extensión	reinforcing steel	acero para refuerzo
face shield	escudo de cara	retardant	retardador
file	limar	road base	base de pavimiento
flag stake	estaca de bandera	rust	óxido
flat ties	ligaduras planas	screed rod	formas para pieza plana
flat work	pieza plana	screeds	botas de goma
float	plana/talocha	setting time	tiempo de fraguado/estab-lecido establecido is more preferred
fly ash	cenizas volantes		
footing	zapata de cimentacíon	shoring	apuntalamiento
footing form	forma de zapata de cimentacíon	slab	losa/plancha
		slab forms	formas de losa
form	forma	sleeve	camisa/manga
form oil	aceite de formas	slump	sección
form ply	madera laminada para formar	smoke	humo
gauges	medidores/metros/gramil	sole plate	placa de base
grade stick	palo/vara de grado	spot footing	zapata de columna
hacksaw	sierra de arco para metal	spreader	huntador
hickey bar	dobladora portatil	stair forms	formas de escalera
hydration	hidratación	stirrup	brida
keel	creyón de madera	striker	percutor
keyway	llave de cimentación		
kicker	pateador		

string line	línea de hilo
tanks	tanques
teamwork	trabajo en equipo
tie off	atar seguramente
tie wire	atadura de Alambre
ties	atadura
tool handles (poles)	mangos de herramientas (tubos)
turnbuckle	tensor
two-by-four	dos por cuatro
two-by-twelve	dos por doce
valves	válvulas
vibrator	vibrador
waler	larguero
waler loops	ligaduras de larguero
wall line	línea de la pared
wall tie	ligadura de pared
water reducer	reductor de agua
weather	tiempo (atmosférico)
wedge	calzo /calzar
welding plate	placa de soldadura
wheelbarrow	senda de concreto carretilla
wire mesh	malla de alambre
wire reel	carrete de alambre
wrench	llave
yards	yardas
Z-clamp	prensa en "Z"

FRAMING / ENMARCADO

FLOOR FRAMING / ENMARCADO DE PISO

backer block	bloque de soporte
backing block	bloque de apoyo
blocking panel	panel bloqueador
bolt	perno/enlace
bridging	puntales de refuerzo
cantilever	viga voladiz
concrete pedestal	pedestal de concreto
dead load	carga muerta
double header	dintel doble
double trimmer joists	viguetas de ajuste doble
double joist	vigueta doble
doubled joists beneath bearing wall	vigueta doble debajo de la pared
exterior wall	pared exterior
first joist	primera vigueta
flange	reborde / pesteña
floor joist	viguetas de entrepiso
I-beam	viga doble "T" "T" or "I"
joist	vigueta

joist bridging	arriostramiento de vigueta vigueta de arriostramiento
joist hanger	soporte de vigueta
lapped joists	vigueta traslapada
length	longitud
live loads	cargas vivas
loads	cargas
plate	plato/viga horizontal
rim beam	viga de borde
rim joist	vigueta de borde
spacing for joists	espacio para las viguetas
stagger-end joists	vigetas para andamio
stairway rough opening	apertura áspera de la escalera
steel column	columna de metal/acero
steel post	poste de metal/acero
stud	montantes
stud spacing	montantes de separación
subfloor	subsuelo
tongue and groove	machihembrado
tail joists	vigueta trasera/cola
temporary braces	frenos temporales
trimmer	condensador de ajuste
web stiffeners	pieza de refuerzo tipo red (web)
width	anchura

WALL FRAMING / ENMARCADO DE PARED

½" spacer	espaciador ½"
1 × 4 braces	frenos 1 × 4
16d nails	clavos 16d
2 × 4 stud	montantes de 2 × 4
2 × 4 plate	placa 2 × 4
2 × 6 plate	placa 2 × 6
2 × 8 plate	placa 2 × 8
4' × 8' sheathing	entablado 4' × 8'
bottom plate	plato bajo
built-up header	cubierta de dintel compuesta
cabinet soffit	sofito de gabinete
ceiling covering	cubierta del cielo
ceiling joists	vigueta del cielo
center to center	centro a centro
centerline	línea central
cripple stud	montantes cojos
double sill	travesaño doble
double top plate	placa superior doble
end stud of partition	montantes finales de partición
exterior	exterior
exterior wall	pared exterior

exterior wall plate	plato de pared exterior
fire blocking	bloqueador anti-incendios
fire stop	antifuego
framing to support lavatory	enmarcado para apoyar el lavatorio/baño
gauge block	bloque de gramil
header	dintel/viga de cabecera
lavatory drain	drenaje del lavatorio/baño
load-bearing interior wall	pared de soporte interior
outside corner post	poste exterior de esquina
partition wall plate	placa de partición de pared
perspective view	vista perspectiva
plan	plan
plan view	vista del plan
plumb	plomada
plumb bob	plomo
plumb line	línea del plomo
sheathing	entablado
sill	umbral / alfeizar
single sill	umbral individual / alfeizar individual
soil stack	bajante sanitaria
sole plate	plato de base
solid header	dintel sólido
string line	línea de travesilla
stringer	travesaño
stud wall	muro con montantes
studs	montantes/barrotes
subfloor	subsuelo
top plate	plato superior
trimmer stud	montante para moldeador
wall	pared
wall brace	apoyo de pared
wall sheathing	entablado de pared
wall stud	montantes de pared
wall tie	pared de amarre

ROOF FRAMING / ENMARCADO DE TECHO	
actual rafter length	largo actual de la viga
backing the hip	soporte/respaldo de esquina
bird's mouth	pico de pajaro
bird's mouth cut	corte pico de pajaro
blade	hoja
body	cuerpo
building line	línea de construcción
ceiling joist	vigueta de cielo
chimney	chimenea
chimney span	chimenea abarbetado
collar tie	vigeta de amarre
common rafter	viga común**

common rafter with overhang	viga común con voladiza colgante**
cripple common rafter	viga común corta
cripple jack rafter	viga de levantamiento corta
doghouse dormer	buharda de casa de perro
dormer	buharda
dormer corner post	poste de esquina para buharda
dormer rafter plate	plato de viga para buharda
dormer side stud	montante de lado para buharda
dormer valley jack	viga de lima de hoya para buharda
double common rafter	viga común doble
double header	dintel doble
double valley rafter	viga de doble lima de hoya
dropping the hip	caida de esquina
Dutch hip roof	esquina de techo holandés
end wall	final de la pared
exterior wall	muro exterior
face-nail from back	cara del clavo desde atrás
face-nail to joist	cara del clavo hacia la viga
flat roof	techo plano
framing square	escuadra de enmarcado
gable	dos aguas/hastial
gable and valley roof	hastial y techo de lima de hoya
gable rafter	viga de hastial
gable roof	techo a dos aguas
gable roof and dormer	techo a dos aguas y buharda
gable roof with shed roof addition	techo a dos aguas con adición de techo cubierta
gable-end stud	vigueta de hastial
hip	lima/lima de hoya/esquina
hip and valley roof	techo de esquina/lima de hoya
hip jack rafter	vigas esquineras
hip rafter	vigas de lima/esquina
hip roof	techo a cuarto aguas
hip roof slopes	techo a cuatro aguas con declives
jack	gato
jack rafter	viga de soporte
lower header	dintel de abajo(door frames)
low-slope (flat) roof	techo (plano) en pendiente-bajo
low-slope roof	techo en pendiente-bajo
main-roof valley jack	techo principal de lima de hoya
mansard roof	techo de mansarda
measuring line	línea de medir
overhang	voladizo/colgante
pitch	pendiente del techo

plate	plato/viga horizontal
primary rafter	viga primaria
projection	proyección
rafter	viga
rafter plate	viga horizontal
ridge	cresta
ridge board	tabla de cumbrera
roof	techo/azotea
roof covering	revestimiento de techo/ cubierta de techo
roof deck	cubierta de techo
roof joist	vigueta de techo
roof sheathing	entablado de techo/ entarimado de tejado
secondary rafter	viga secundaria
shed roof	techo de cobertizo
side cut	corte de lado
slope	cuesta/inclinación/pendiente/ declive
span	abarbetado/luz/vano/claro
tail	cola
theoretical rafter length	largo teórica de la viga
toenail	clavo oblícuo
toenail joist	clavo oblícuo para vigueta
tongue and groove	machihembrado
top plate	plato superior
total rise	elevación total
total run of overhang	corredor total de voladizo/ colgante
unit rise	unidad elevada
unit run	unidad corrediza
upper header	dintel superior
valley jack rafter	cabio de lima hoya
valley rafter	viga de lima hoya
valley strip	tira de lima de hoya

TRUSSES / CERCHAS/CABALLETES <PART OF ROOF FRAMING>	
2 × 4 blocking	trabas/bloques 2 × 4
2 × 6 backing	soporte 2 × 6
3-hinged arch	arco con 3 bisagras
attic	atico
bearing wall	murallas de soporte
bottom cord	cuerda baja
camber	comba
cambered	combado
cantilever	voladizo
connector plate	plato de conector
cord	cuerda
fan	ventilador
gable end	hastial

hip	lima
inverted	invertido
king post	poste principal
length	largo
mansard	masarda
multipanel	panel multiple
overhang	voladizo/vuelo/alero/colgante
partition clip	clip de partción
partition wall	pared de partición
scissors	tijeras
slope	inclinación/pendiente/declive
span	abarbetado
split-ring connector	conector de anillo de partición
system 42	sistema 42
vault	bóveda
web	red

STAIR PARTS / PARTES DE LOS ESCALONES	
angle newel	barandilla de ángulo
balcony	balcon
baluster	barandilla
banister	pasamanos
baseboard	zócalo/moldura base
bullnose	canto biselado
cap	tapa
cove mold	moldura ensenada
face	frente
fascia	fascia/faja/imposta
filled	llenado
fitting	accesorios
flared tread	huella acampanada
floor	piso
flooring	revestimiento para pisos
gooseneck	cuello de ganzo
handrail	pasamanos
inner string	cuerda/hilo interior
intermediate	intermedio
landing	descanso
landing tread	descanso de peldaño/huella
moulding trim	moldura decorativa
moulding	moldura
newel post	poste de escalera de caracol
nosing	vuelo/orilla del escalón
nosing projection	protección de vuelo/(en la orilla del escalón)
outer string	cuerda/hilo exterior
rail	riel
rail drop	caida de riel
rake	rastrillo
return nosing	moldura de retorno

riser	contrahuella
skirt board	falda
skirting board	tabla para falda
starting riser	contrahuella principal
starting step	grada principal
stringer	hilada/cordelada
tread	huella/peldaño (escalón)
tread bracket	ménsula de peldaño
tread cap	caperuza de peldaño
tread projection	protección de peldaño
turnout	sacar/salir
up railings	rieles hacia arriba
wall rail	riel de pared
wall string	hilo de pared